Much Ado About *Almost* Nothing

Man's Encounter with the Electron

Hans Camenzind

ISBN-13:   978-0-615-13995-1

Edition 1.1
June 2007

BookLocker.com

Contents

The author is indebted to Robert A. Pease and Reinhard Zimmerli for proofreading this book.

1. Almost Nothing

Ancient superstitions, a skeptical Greek,
a siege in Italy, a physician in London
and shocking entertainment.

The little thing that threads through this book truly amounts to almost nothing. It is so small you can't see it, not even with the most powerful microscope. It is elusive, as if it were shy: it can appear as a tiny point with a weight (or more properly a mass) of much less than a billionth of a billionth of a billionth of a gram, scurrying around at near the speed of light; or it can shed its body and become a ghost, a wave.

We would never have noticed this strange particle were it not for the fact that there are many of them. They are everywhere, in every material, in numbers so huge they are beyond any human comprehension. In the point at the end of this sentence there are some 20 billion of them.

It was only a little over 100 years ago that man began to deduce the nature of this weird, minuscule thing and gave it a name: the electron. For most of history he was unaware of its existence, ascribing its effects to various gods. A large, powerful god who hurled lighting bolts, such as Zeus, Thor, or Jupiter. Or little ones who hid in amber, which attracted feathers or straw when rubbed, or in the lodestone (magnet), which attracted other stones of the same kind.

We shall follow man's encounter with the electron in this book: how a few people overcame their superstitions and began to investigate; how the electron gradually became useful, though man still had no idea what it was; how it finally revealed itself and then grew in importance to such an extent that we are now completely, utterly dependent on it.

The first incident of note happened around **600 BC**, in the city of Miletus (or, more accurately, Miletos). The city was Greek, but it was located on what is now the west coast of Turkey. At that time Miletus was the richest city in the Greek world and the most accomplished cultural center. Part of its status was due to Thales, who had gained fame in mathematics and astronomy and also proved himself to be a shrewd businessman [2].

Miletus had gained its wealth through trade. It had an excellent harbor and was strategically located on the way to Egypt from the north. And the trade routes already extended all the way to the Baltic sea, 2300km (1400 miles) away, from where the traders brought the most valuable gem then in existence: amber. Amber is a resin from trees, which grew some 40 million years ago, and are now deep underground [8]. Over time the resin has hardened and now appears to be a bright yellow rock; because of its sun-like appearance the Greeks named it ηλεκτρου (elektron).

Unlike other gems, amber has a fascinating property: when rubbed, it attracts light objects, such as straw. The Greeks (and just about everybody else) believed there was a tiny god inside. That was the explanation for everything they couldn't understand. The Greeks had a very large number of gods: every object on earth or in the heavens had a god, everything good or bad was personified as a deity, usually in human form; and each god had its own fantastic story.

There was another effect that couldn't be explained: magnetism. In some regions near Greece one could find rocks, which attracted each other (called a lodestone). More gods were created.

As far as we can tell, Thales was the first person to debunk this belief. He investigated the lodestone and amber and pronounced that the effects were not due to gods. But it was only a small step forward: his explanation was that amber and the lodestone had a "soul". (He also believed that the earth was a flat disk floating on water, and that everything comes from water and returns to water).

But we have to keep in mind that we know very little about Thales, about what he thought and said. Our only source of information is hearsay, comments by later scholars. If he wrote anything, nothing survived.

When you had something important to say in 600 BC, you wrote it on a scroll and let other people read it. There were no public schools, universities or libraries, not even in Miletus; only a small minority of people could even read. If your dissertation was popular, a rich, educated collector had a copy made by a scribe; if your ideas were out of favor, your scroll eventually decayed and was discarded.

Shortly after Thales, Miletus started to decline and Athens became the center of Greece. Miletus was captured by the Persians, its harbor silted up and then the town disappeared altogether.

For more than 1800 years there was only silence on the subject. It appears that no other Greek (or Roman) picked up where Thales left off. If there were any, all traces of them have disappeared.

And when the Roman Empire became Christian there was no longer any progress in other fields of science either. The church taught that every object and every effect was controlled by divine interference and thus the investigation of nature not only made no sense, but was sacrilegious. This belief was enforced by the inquisition.

So it is a surprise that we find a rather detailed investigation of the magnet during this time. It is in the form of a letter written in Latin by a Pierre de Maricourt to his neighbor [10]. We know very little about the author, who signed the letter as Petrus Peregrinus (Peter the Crusader); Roger Bacon mentions him in his writings and gives us the impression that Maricourt was one of the most impressive and knowledgeable people he knew. At the bottom of the letter it says: "Finished in the camp of the siege of Lucera on 8 August **1269**".

What was a Frenchman doing in Italy, laying siege to a town? Lucera is a town 200km (130 miles) east of Rome (not in Sicily, as most historians state) [9], and had become populated by Muslims. The pope took offense and asked the French for help. An army was dispatched, led by the brother of the king; Maricourt was almost certainly an engineer who built and operated the catapults, which hurled rocks and flaming tar-balls into the town.

A siege tends to be boring for the aggressor; one has to wait until the population starves. Thus Maricourt had plenty of free time on his hands and he put on paper what had been his passion for some time: investigating the magnet. He had found that each magnet has a south pole and north pole, which he determines by carefully tracing the magnetic field with an iron needle. With two magnets the north pole of one attracted the south pole of the other, but when two north or south-poles where held together they repelled each other. When he cut a magnet in half, each piece had its own north and south pole. He also found that he could magnetize a piece of iron with a lodestone and that a strong lodestone was able to reverse the magnetism of a weaker one.

Maricourt suggested improvement for the compass, which had been in crude use by mariners: a 360-degree scale which would let you find the course to be steered.

But he made two grave mistakes. He believed that the magnet in the compass pointed toward the north star, not the north pole of the earth. And, at the end of the letter, he proposes a motor with magnets that would run forever, a perpetuum mobile. If it doesn't work it is probably due to the lack of skill by the one who is building it, he said. The motor never had a chance of working.

That is all we know about Pierre de Maricourt. His letter was copied occasionally; some 28 copies are still in existence. No two copies are the same, all were altered while copying, the majority of them substantially. Which goes to show how uncertain our knowledge of the ancient past is.

For the next 300 years progress in the understanding of either magnetism or electricity again slowed to a crawl until in England, at the end of the 16th century, we find a physician by the name of William Gilbert who, as a hobby, investigated the phenomena of the lodestone and amber. In addition to medicine he had studied physics and astronomy and, while establishing a prosperous medical practice in London, he read everything available on the effects of the lodestone, the compass and amber. He interrogated countless sailors on their experience with the compass. But, above all, he experimented. Taking little for granted, he tried to verify every statement and belief[6]; it took him 18 years.

William Gilbert

In **1600** he published his findings in a book called De Magnete [3]. It became an instant bestseller, as bestsellers went in those days. In it he described the entire history of the beliefs surrounding the lodestone and amber and exposed dozens of misconceptions and superstitions. Garlic, for example, was supposed to destroy magnetism, so sailors were not allowed to chew it around a compass. Gilbert smeared garlic all over magnets and found no effect. The myth appears to have started with a bad copy of a book by Pliny (about 75 AD) where the word alio (other) was misspelled allio (garlic).

Gilbert discovered that lodestone and iron are the same and that iron can acquire poles if magnetized by a lodestone. He found that the magnetic effect penetrates paper or linen and worked even under water, whereas the effect of amber ceased in moisture. He vividly described how amber was not the only material that could cause attraction when rubbed, but that diamond, sapphire, opal, amethyst, rock crystal, sulphur, glass and many other materials did the same. He separated all known substances into "electrics" (which could attract when rubbed) and "non-electrics"[6].

His main interest, though, was in the magnet; he mentions electrics only briefly. He found that the earth is a magnet and that the compass points toward the North Pole, not the North Star. He explained the dip in the compass needle, described the variation of the magnetic field over the world and determined that the magnetic poles of the earth are shifted slightly from the geographic poles.

Gilbert made a model of the magnetic properties of the earth by filing a lodestone into a sphere. By doing this he could visualize and demonstrate much, but it also led him somewhat astray. This little sphere looked and behaved so much like the earth and he was so enamored by magnetism, he came to the conclusion that objects (like people) were held to the earth by magnetism. He also missed the fact that magnets (and "electrics") can not only attract but also repel.

But with all his studying and investigation he still did not know what either effect really was. He viewed magnetic attraction as entirely imponderable and incorporeal. He speculated that the "electrics" emitted "a subtle diffusion of humor or effluvium."

The year De Magnete was published, Gilbert became president on the College of Physicians and in 1601 physician to Queen Elizabeth. He died two years later, at age 59.

The next place we visit is Magdeburg, Germany. The time: around 1666. Otto Guericke (who was later raised to the peerage by the emperor and thus called *von* Guericke), had been the town engineer, then became a politician and was now the mayor of the city. He was also the town brewer, which made him rich.

Guericke had been an experimenter for most of his life and had acquired fame with his studies of vacuum. He had devised a pump similar to the ones used for water, but with a piston so tightly fitted that it could pump the air out

of a hollow metal sphere. He demonstrated that you could not hear a ringing bell inside the sphere, i.e. that sound does not travel through vacuum. Then he built two hollow hemispheres, held them together, pumped the air out and showed that the two halves could not be pulled apart by two teams of eight horses. He repeated this brilliant piece of showmanship several times before kings and emperors in Regensburg, Vienna and Berlin [6]. (Actually, the second team of horses did nothing. He could have tied the chain to a large tree or building and gotten the same force, but two teams of horses were certainly more impressive).

Otto von Guericke

But now he had moved on to a different field: electricity. He poured a mixture of molten sulphur and minerals into a glass sphere. When the sulphur had cooled and become solid he broke the glass vessel, drilled a hole through the sulphur ball and put a rod through it. Then he mounted a crank at the end of the rod and put the assembly into a wooden stand. When he turned the crank and held his hand against the rotating sulphur, he could draw sparks from it [5].

Gilbert had favored magnetism because the effect was much stronger. Guericke increased the effect of the "electrics" manifold.

After spinning the sulphur and mineral globe and touching it with his hand, it would attract down feathers and other light objects over a much greater distance, and he noticed something very unusual. At first the feather was attracted to the globe. But when it touched the globe, the feather suddenly fled from it. The most curious behavior appeared when the charged globe was close to the floor and a feather was in between. In this situation the feather would dance between the ball and the floor. At first the feather was attracted by the charged sulphur ball. As soon as it touched it, though, it would move away from the ball toward the floor. Then, as it touched the floor, it would be attracted by the ball again.

Guericke then built larger contraptions with belts and wheels, which spun larger globes faster and created larger sparks. In **1672** he published the results of his experiments [5]. He confessed that he didn't understand the reason for the feather's strange behavior. "It must be the globe's own decision", he writes "whether it shall attract or repel. When it does not want to attract, it doesn't attract; nor does it permit the feather to approach it until it has cast it against something else, perhaps in order to acquire something from it."

Guericke made two more startling observations. When he held the globe to his ear, he heard "roaring and crashing." And when he attached a linen thread to the globe "an ell long" (a little more than a meter) he noticed that the linen thread also attracted light objects.

In **1729**, 62-year-old Stephen Gray made the next discovery. He had entered the Charterhouse, an institution for gentlemen who had fallen on hard times, ten years before. Although a dyer by profession, Gray had been involved in science and loved to experiment with electricity. Apparently he hadn't heard of Guericke or couldn't afford to build one of his machines. To generate electricity he used a glass tube, which was plugged with a cork at either end "to keep the dust out". When he rubbed the glass tube, he noticed that a feather was attracted not only to the glass, but to the cork as well. He concluded the cork must conduct the "electric virtue", and decided to test other materials. He drove a wooden stick into the cork and fixed an ivory ball at the end of the stick. When he rubbed the glass tube, feathers were attracted to the stick all the way out to the ivory ball. He replaced the wooden stick with a metal rod, and that worked, too [4].

Becoming excited now he wound twine around the glass tube and hung the twine down 10 meters (30 feet) from the balcony of the Charterhouse. Near the ground he affixed his ivory ball. And, sure enough, when an assistant rubbed the glass tube above on the balcony, Gray observed that feathers were attracted to the ball. He replaced the ivory ball with everything he could find around the Charterhouse without spending any money: stones, tile, chalk, coins, a kettle from the fireplace, vegetables and plants from the garden. They all worked.

Gray now ran out of height; the balcony was as high as he could go. Fortunately he met one Granville Wheler, who had a large house. Together they strung twine around Wheler's house, suspended by short lengths of the same twine. But this time the ivory ball at the end didn't attract anything.

The two amateur scientists suspected that it must be the way they suspended the twine, so they changed the suspending strings to thin pieces of silk. Now the ivory ball was as alive as before and they managed to conduct electricity over a distance of 100 meters (300 feet) [6].

But as they attempted longer distances, the silk threads began to break. Let's use something stronger that's also thin, they said to themselves. So they tried thin pieces of brass wire. But that didn't work at all.

It must be the material then, they thought, not its thickness. They used silk to suspend the twine again, this time thicker and stronger and achieved a distance of 253 meters (760 feet). Within a short time they learned which materials make good "insulators": silk, horsehair, resin and glass. Suddenly Gilbert's distinction between electrics and non-electrics made no sense anymore. The important distinction was now between insulators and conductors.

For his discovery Gray was made a member of the Royal Society and received a medal. He died at age 70, considerably more distinguished but still penniless.

In **1733** Charles Francois Dufay learned about Gray's experiments. The 35-year-old Dufay had been a soldier. His main interests were in chemistry, anatomy, mechanics and botany and he had already contributed a number of original papers to the French Academy [6].

Dufay absorbed the new subject of electricity rapidly. He experimented and found that twine conducted best when wet and that metals were better conductors. With this new knowledge he extended the distance of electrical conduction to 420 meters (1300 ft).

To determine the presence of electricity, Dufay used a thin gold leaf (instead of a feather) and he found something that had been missed until then. Some materials when rubbed attracted the gold leaf, others repelled it. Aha!, he said, there are two kinds of electricity, not just one. He called one kind vitreous (for glass and like materials), the other resinous (for resin).

Dufay was also the first one to notice the effects of electricity on the skin. He described how electricity passing over his face felt like a spider web and how it sometimes could sting and burn when discharged through his fingers [6].

That was a mild foretaste of what was to come. In the span of only five years electricity was promoted from an obscure phenomenon to the most fashionable "science". What took place was a veritable mania to build "electrical machines", ever larger glass globes, no longer rubbed by hand but by small leather cushions. The faster these globes or disks were turned, the more electricity they produced, so they were spun with the help of ever larger wheels and belts

The movement started in Germany and one of the busiest demonstrators during this period was Georg Mathias Bose, a professor at Wittenberg. Bose was a born showman. In one of his famous demonstrations he insulated an entire dinner table and connected it to an electrical machine hidden in the next room. After the guests had sat down he gave a signal to his assistant who operated the machine with all his force. The surprised guests had sparks flying at them as they touched the silverware and the metal dishes and cups [6].

As a climax he placed a "pretty young maiden" on an insulated stand, connected her to the machine and invited the young men among his guests to kiss her. Some of the brash young kissers thought they had their teeth knocked out (no mention is made of the teeth of the young lady). Professor Bose published his experiments in three small volumes of Latin poetry, strange pieces of literature reeking of self-admiration [1].

Bose also administered a shock to 20 soldiers holding hands and later bragged that he could do this to an entire army.

Bose and his contemporaries generated sparks of considerable length, which we now know, corresponds to voltages in excess of 50,000 volts. This is far above the 110 or 220 volts in houses, which is quite capable of killing a person. So why weren't these people killed? They were lucky that these electrical friction machines were very inefficient power generators; they produced a high voltage alright but, as soon as this voltage was applied to the human body it collapsed to a harmless level.

Pieter van
Musschenbroek

In **1746** Pieter van Musschenbroek, a professor of experimental physics at Leyden (or Leiden) in Holland, tried to draw "electrical fire" from water in a glass jar. Musschenbroek had an acquaintance, a lawyer named Andreas Cunnaeus, who amused himself by visiting Musschenbroek's laboratory and dabbling in electrical experiments. Cunnaeus tried the same experiment and, when he grabbed the outside of the jar with one hand and touched the conductor immersed in the water with the other, he received a tremendous shock. He showed the experiment to Musschenbroek, who repeated it with a larger jar. The shock Musschenbroek received was so stunning that he thought he "was done for".

Hours later he was still shaken and vowed that nothing could make him try the experiment again [6]. For anyone who thinks that lawyers have never contributed anything to mankind, here is a clear-cut case of at least one lawyer who did. It is ironic that this case, too, was a painful experience.

Musschenbroek sent a letter explaining his experiments to his correspondent at the French Academy and soon the "Leyden Jar" was in fashion, with metal foil around its outside replacing the hand. Every electrical experimenter worth his salt built one and started shocking everybody in sight. Among the experiences they reported were nose bleeding, temporary paralysis, concussions and convulsions. One gallant experimenter observed that his wife was unable to walk after he had asked her to touch the Leyden jar.

Musschenbroek believed that the electricity was in the water inside the jar, and that the larger the volume of the water the greater the shock. What he or Cunnaeus had in fact discovered was what is now called the capacitor. The electricity (or better: electrical charge) is temporarily stored not in the

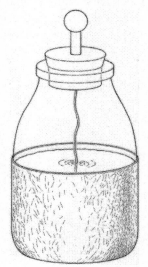

The Leyden Jar: It stored electricity, which made the shock almost lethal. But nobody knew why or how.

water but in the *wall*. The electrons generated by rubbing of the glass accumulate on one side of the wall or the other and wait there until they find a path to flow out of the capacitor again (e.g. through the human body), in great strength and all at once. In this way the delivered power becomes larger, and shocking people more fun.

And none provided more fun than Abbé Jean Antoine Nollet. He was not really an Abbé, though. He had studied theology but, before becoming a priest, switched to physics. Nevertheless he wore the garb of a priest and always used the title.

Nollet had connections at court. Louis XV invited him to demonstrate this new electric marvel at Versailles. Nollet had 180 of the King's soldiers join hands in a circle and then applied the Leyden jar to one of them. To the great joy of the king and his court the soldiers simultaneously leaped into the air. Louis liked the demonstration so much that he had it repeated in Paris, this time with 700 Carthusian monks [1].

These frivolous demonstrations depict the state of electricity in 1747. Electricity had become a more interesting effect than magnetism, but both were still only on the level of parlor tricks; with an electric demonstration anyone could become the life of a party.

2. Benjamin Franklin, Electrician

After a successful career as a business man
he dabbled in electricity with three of his friends
and turned out to be a first-rate scientists.

In 1747 America was a remote outpost, a rough, agricultural country.
As every aspiring experimenter knew, things of any scientific importance
happened in France, England, Germany or Italy - perhaps occasionally in
Spain or Holland, but certainly not in America. But they hadn't counted on
Benjamin Franklin.

Benjamin Franklin's father had moved to Boston from England. He was
a candle and soap maker and ran a shop opposite the Old South Church on
Milk Street. Benjamin, the youngest boy, was born in 1706.

Boston in those days was a village with some 6,000 inhabitants and open
sewers ran through the middle of the dirt streets. At age eight Benjamin went
to school, but stayed for only two years. He was apprenticed to his older
brother James, whose printing shop was doing moderately well and who soon
started to publish a somewhat subversive newspaper. The relationship
between the two brothers grew strained, especially when James, because of
his attacks on the government, was forbidden to publish and the paper had to
be continued under Benjamin's name. In 1723, at age 17, Benjamin walked
out. To make sure he couldn't find a job, James blacklisted him with the other
printers in Boston [16].

Franklin decided to find a job in New York, at that time a village of
some 4,000 inhabitants (while Boston had grown to 12,000). But there was
only one printer and no work for Franklin. Someone thought that there might
be work in Philadelphia, the second largest city in the colonies after Boston.
So Franklin traveled on, by boat, to Philadelphia. He arrived nearly out of
money, but he found a job in the printing shop of Samuel Keimer.

Apart from being close to literature, working in a printing shop had
another advantage for a bright young man in those days: since it produced the
main means of communications - handbills, books, pamphlets, government
proclamations - it was frequented by the merchants, the educated elite and

government officials. It was in this way that Franklin met Governor Keith.

Keith was impressed by the young man who was well read and full of ideas. He suggested to Franklin that he should go to London, acquire printing equipment and start his own shop. Keith promised to sponsor the trip, but he was the type of man who promises much and delivers little. Franklin quit his job and packed his things. He boarded the ship, expecting Keith's letter of recommendation and money to be delivered on board before sailing. But the ship sailed without anything from Keith - and with Franklin on board. He arrived in London, broke again [16].

He found a job in a large printing establishment and was able to support himself fairly well. The sophisticated environment added to his skills as a printer - and got him into trouble. Despite the maxims, which are so often ascribed to Franklin, he was a freethinking, free-loving young man. He was torn between a strong sexual drive and a strict moral code, which he imposed on himself and practiced seriously. He would write out each of the desirable virtues (12 of them), practice each of them for a week, and then promptly squander his hard-earned money on prostitutes [13].

After two years in London, a Philadelphia merchant made Franklin an offer to pay passage back to Philadelphia if he would become a clerk in his store. There was a possibility that Franklin might become a partner in the store. He accepted and returned to Philadelphia, but after only a few weeks the merchant died and Franklin was out of a job again.

Samuel Keimer took him back. The 21-year-old Franklin had become a highly skilled printer and first-class writer, but he was itching to become independent. After only six months he quit Keimer's employ and set up a printing shop with a partner, whose father provided the money. The partnership hovered near bankruptcy for the next four years. In the meantime, Franklin's way of life was catching up with him: he had fathered an illegitimate child. To provide a home for his child, Franklin took a common-law wife and, after buying out his partner with borrowed money, he opened a shop where he sold ink, paper, books and anything else people were willing to buy.

Now that he had a family, Franklin settled down to hard work. His wife turned out to be the ideal business partner. A stout woman, she kept the books (although she could barely spell), managed the store and raised the children. Besides the illegitimate son, the Franklins had a son of their own (who died at age four) and daughter.

Within two years the debt was paid off and the Franklins were on their way to success. Keimer was out of business and Franklin took over a small newspaper from him, the Pennsylvania Gazette. With Franklin's writing skill and humorous style, the paper quickly went from a money loser to a profitable business venture. Shortly after bringing out his newspaper Franklin added "Poor Richard's Almanack" to his business ventures.

Franklin was now a successful businessman. He joined the Freemasons and became a Warden, then Grand Warden, then Secretary and finally Grand Master of the Province.

When he started his newspaper there were problems distributing it. A rival printer who also issued a newspaper held the office of postmaster. This printer had the privilege to distribute his own newspaper free of charge but in addition he ordered his riders not to carry any others. Franklin used his connections to take the appointment away from the rival and become postmaster himself. The post office of Philadelphia was now in his store and the postal riders carried all the newspapers [16].

But it was not all business and social status. Franklin was an avid reader and was genuinely interested in almost everything he read in books and newspapers. He had founded the Junto, a men's club which met once a week and discussed such subjects as science, politics and religion. On Franklin's urging the Junto sponsored the first public library in the colonies, the Philadelphia Library Company. Each member made an initial and then yearly contribution. The Library Company imported books from England and made them available to everyone. Franklin was also the organizer of the Union Fire Company, whose members made themselves leather buckets and held them ready at their houses for emergencies.

The more successful Franklin became, the more time he spent on social and political projects. In 1736 he was made Clerk of the Pennsylvania Assembly (which also brought him the printing business of the Assembly). In 1744 he originated the American Philosophical Society, an outgrowth of the Junto.

Besides his interests in social and political questions, he was intrigued by science. He devoured all the articles and books on science the Library Company received. And he was an eminently practical man who knew how to use tools and fashion things.

Thus, at age 40 we find Franklin a highly successful businessman with the beginnings of a political career. He owned a general store, a printing shop, a newspaper and an almanac outright and he was a silent partner in a printing shop and newspaper in South Carolina and New York. But Franklin was not entirely happy with this situation. Making money was beginning to bore him; he needed a change. What he really wanted to do was to devote his entire time to science. He had what we would call today a mid-life crisis.

In 1748 Franklin gave a half-interest in his print shop to his foreman and became a silent partner. From that point on and for the next 18 years he received £ 500 per year from the print shop. He moved his family to a new house at the edge of town (a few blocks away, that is) where he would be less disturbed and he began to experiment with electricity [2].

Why did he choose electricity? It was the newest and most attractive of the sciences. It had just become the rage in Europe and was now spreading to the American colonies. Franklin had heard the lecture of a Dr. Archibald Spencer on a visit to Boston in 1743. Spencer was an itinerant (and not very successful) lecturer from Scotland. Franklin organized a lecture for Spencer in Philadelphia and later bought his equipment [5].

But the incentive seems to have come from Peter Collinson, Franklin's correspondent in London for the Library Company. Collinson was a cloth merchant who was also interested in science, specifically in botany. He belonged to the Royal Society and had a magnificent botanical garden in London. He purchased books for the Library Company and kept Franklin informed on the latest scientific happenings in Europe. In one of the shipments of books he included an issue of the "Gentleman's Magazine," which had an article about electricity, together with a glass tube for experimentation.

With three of his friends from the Junto, Philip Syng, a silversmith, Thomas Hopkinson, a lawyer and Ebenezer Kinnersley [12], an unemployed preacher, Franklin began to experiment with electricity. To any European savant it would have been a ridiculous picture, a printer, a silversmith, a lawyer and a preacher setting out to move to the forefront of science.

But the difference was the mind of Benjamin Franklin. He went about the investigation in a deliberate, determined way. He had the advantage of knowing practically nothing about electricity when he started out and so was not hindered by false assumptions and theories.

In their first experiment the motley foursome used a small cannonball, which was put on an insulated stand. A ball of cork was hung on a silk thread so that it touched the cannonball. When they charged the cannonball with the rubbed glass tube, the cork ball moved away from the cannonball, indicating the presence of an electric charge. So far they had only been repeating old experiments, but Hopkinson discovered an interesting effect. When he tried to discharge the cannonball with a sharply pointed conductor, he noticed that the cannonball lost its charge before the needle actually touched it. In other words it wasn't even necessary to make contact with the charged object if the discharging conductor had a sharp point. At night they also noticed a faint glow on the tip of the point [2]. They were puzzled by this and wrote it down.

The next experiment was Franklin's. He placed two of his friends on stands insulated by wax. The first one rubbed the glass tube and held it to the second one, who drew a spark with his knuckle. Franklin discovered that a third man, standing on the un-insulated floor now could receive a shock by touching either of the two men on the insulated stands, but the shock wasn't as strong as the two men on the stands had felt

Franklin's charge experiment

between them. If the two men on the insulated stands held knuckle to knuckle immediately after charging themselves with the glass tube, they would draw another spark and the man standing on the floor subsequently would feel nothing [2].

Franklin now had one of those brilliant insights that are so hard to explain and are, therefore, ascribed to genius. He theorized that electricity was not created by the glass tube as all the scientists had assumed, but that rubbing the glass tube only transported it. It must be, he thought, that after the first person rubbed the glass tube and held it to the second on the insulated stand, the second person has an excess of electrical "fluid" and the first one a deficiency. When the two touch each other, the electrical particles move back to their original positions. If a third man, standing on the floor touches either one of the men on the stands, he receives only half the charge. He pictured an excess of electrical particles as being positive and a deficiency as being negative. In a discharged body the number of electrical particles, he conjectured, was just right and thus it is electrically neutral. Therefore, there were not two electrical fluids, but only one type of electricity. This was in direct contradiction to the accepted theory [10].

It was a brilliant deduction, but Franklin got the polarity wrong. He couldn't have known that the electrical "fluid", the electron, has a *negative* charge, not a positive one as he assumed. Since opposite polarities attract, the electrons flow toward the positive terminal. Ever since Franklin we think of current flowing from the positive terminal of a battery to the negative one, while in fact it is flowing in the opposite direction.

Philip Syng, the silversmith, soon got tired of rubbing the glass tube and reinvented the glass globe with the crank. The group was apparently not aware that such globes had been in use in Europe for years [10].

Franklin wrote a letter on these electrical experiments to Collinson and hoped that he would use his influence with the Royal Society to get it published. He didn't know how much of it was new and how much was old, so he was cautious in his wording. Collinson showed the letter to his acquaintances at the Royal Society and it was read before a meeting. It clearly contained some startling new discoveries and Franklin's explanation was intriguing; the letter was published in the Philosophical Transactions [2].

Shortly after sending the first letter, Franklin heard about the Leyden jar (he called it Dr. Musschenbroek's wonderful bottle) and decided to investigate it. He found an immediate confirmation of his theory. Together with Kinnersley, the preacher, he discovered that a Leyden jar could be charged backwards by grounding the inside and charging the outside. In this way the outside became positive, the inside negative.

The next step was a logical deduction: if his theory was correct, then several Leyden jars could be charged separately and then connected in series, with the outside of one jar connected to the inside of the next. This should make the electrical shock much stronger. It did; it made the shock so strong that he nearly killed himself.

The exuberant group now tried to find a practical application for their discoveries. They killed a turkey with an electric shock and swore that it made the meat more tender. They investigated the possibility of curing paralysis by electric shock, but found that it had no lasting effect [13].

But what made the Leyden jar really work? Where did it store the electricity? Franklin was determined to find out and devised a surprisingly simple experiment. He charged a jar, removed the cork and the wire (connecting the water through the cork to the outside) with an insulated tool and put the cork and wire from an uncharged jar into the bottle. The jar delivered a shock as before; therefore the charge was not in either the wire or the cork.

Next he charged the jar again, removed the cork and wire and decanted the water into a second bottle. He placed the cork and wire into the second bottle with water in it and received no shock. He then filled the old jar with fresh water, put a new cork and wire into its mouth and was able to get a shock. This meant that the electric charge must be in the glass [10].

Did the shape of the jar have anything to do with the effect? He made a sandwich with two lead plates and a pane of glass in the middle and was able to charge it as if it were a Leyden jar [10].

He reported his findings in letters to Collinson who saw to it that they were printed in the Philosophical Transactions. He expounded more of his theory: he speculated that the "fluid" consisted of tiny particles and that there were two types, common and electrical. Because of the great subtlety of electrical particles, he thought, they can move through common materials like water moves through a sponge. When the pores of this sponge are filled, common matter is neutral. If they are empty, it is negative. If electrical particles are added to a filled sponge of common matter, they surround the surface of the common matter in an "atmosphere" or cloud [10].

Franklin's theoretical speculations had some flaws, especially with the behavior of negatively charged bodies. According to Franklin's theory they should have attracted each other, but in fact they repelled in exactly the same way as two positively charged bodies did. Franklin became aware of this but never managed to account for it [11].

Franklin moved on. There was one thing, yet to be investigated, that had intrigued him from the beginning: lightning. To almost all electrical experimenters of that time the flash and sound of electrical sparks suggested the similarity with lightning. Franklin felt that clouds were charged by friction similar to a glass globe. But a mere belief was not good enough; it needed to be proven.

He wrote another letter to Collinson, suggesting an experiment: on a high tower or steeple build a sentry box in which a man can stand without getting wet in a thunderstorm. Erect an iron rod, pointed at the end and reaching some 10 meters (30 feet) into the sky. The iron rod rests on an insulated stand. Now, with thunderclouds overhead, the man in the sentry box should be able to hold a grounded wire to the rod drawing

Franklin's proposed sentry-box experiment, which was first performed in France.

sparks identical to the ones drawn from a glass globe or a Leyden jar [4].

Franklin was applying his experience with the needle and the cannonball. If the clouds are indeed charged, a pointed rod close to them should behave the same way, discharging the clouds before the charge built up to such a level as to cause a lightning flash.

In the same letter he also suggested the erection of lightning rods. He was convinced that such a pointed rod, sticking out above the tallest part of the house and connected through a wire to the ground would safely absorb lightning strokes thereby preventing damage to the building [4].

This time the Royal Society balked. They felt that this letter consisted mostly of conjecture, not fit to be printed in the Philosophical Transactions. Collinson thought otherwise. He had Franklin's letters printed in a book entitled "Experiments and Observations on Electricity made at Philadelphia in America, by Benjamin Franklin". The book came out in **1751**, was 86 pages strong, cost two shillings sixpence, and didn't sell [1].

But fate takes strange turns. One copy of the book got into the hands of Comte de Buffon, the naturalist. Buffon disliked Reaumur, who was a close friend and supporter of the Abbé Nollet. Buffon saw an opportunity here: Franklin's theories clashed with the views held by Nollet. So Buffon asked his friend, the botanist Dalibard to translate Franklin's book into French. Dalibard added to the translation a brief history of electricity in which Nollet was entirely ignored [4].

Neither Buffon nor Dalibard had any knowledge of electricity. They hired an experimenter as advisor and together they repeated Franklin's experiments and made sure the king heard about their endeavors. The plot worked. The King asked for a demonstration, which Buffon and Dalibard gladly performed. Louis XV was so enamored by the Franklin experiments that he ordered the three to also perform the last one, the sentry-box experiment.

This Buffon and Dalibard had not planned on. The sentry box experiment had never been performed and failure would be embarrassing. They decided to perform the experiment in secret first. Dalibard set up a sentry box and insulated metal pole at Marly, a small village outside Paris. He hired an old dragoon to stand by and wait for a thunderstorm.

On May 10, 1752 a thunderstorm appeared over Marly. The dragoon tried the experiment as instructed. When he approached the bottom end of the metal pole with the grounded wire, sparks jumped out and he got so frightened that he ran and fetched the village priest, who repeated the experiment and reported it to Dalibard [4].

The news spread rapidly through France and England. The sentry-box experiment was repeated in different places. There was no doubt now that lightning was electricity. Buffon was triumphant, Nollet was furious. His status as court electrician was threatened and he attacked Franklin and his theories. At first he didn't know who Franklin was and thought that he might be a person invented by Buffon to embarrass him. But he soon became aware that Benjamin Franklin was real, a formidable electrician and a man not easily ignored.

Franklin hadn't heard yet of the successful performance of his experiment or the fight that Nollet was putting up. He had proposed the experiment rather than do it himself because there was no tower or high point in Philadelphia. But in the meantime he had thought up a different way to achieve the same result. He built a kite and placed a wire on top of it so it would point upward. The wire was also connected to the kite string which, when wet, would conduct electricity [4].

In June of **1752**, in the middle of a stormy night, he and his 20-year-old son William went to the commons and flew the kite. At the end of the string they had affixed a length of silk ribbon and they stood under a roof so that the silk would not become wet and conduct electricity to their hands [4].

At the junction of the silk ribbon and the string Franklin had attached a key. He approached the key periodically with his knuckle to see if he could draw a spark. But he had to wait for a good cloud overhead and for the string

Benjamin Franklin after he became famous as an experimenter.

to become wet. Suddenly he noticed that the frazzled ends of the string were standing upright. He approached the key with his knuckle and a spark jumped out. He took a Leyden jar and was able to charge it through the key [4].

With this double confirmation of the identity of lightning Franklin's fame was assured. During his lifetime his book on electricity went through five editions in English, three in French, one in German and one in Italian, and he received honorary M.A.'s from Harvard, Yale, and William and Mary, Ph.D.'s from St. Andrews and Oxford and the Copley Gold Medal and Fellowship from the Royal Society [3].

But with the lightning experiment and the invention of the lightning rod Franklin's days as a scientist came abruptly to an end. His scientific period had lasted less than seven years. Politics was beckoning. With his influence as a newspaperman and his fame as a scientist Franklin was an ideal candidate for office. In 1753 he was made Deputy Postmaster-General of North America. He went about the job with his usual vigor. He improved efficiency and abolished the monopolistic system for newspaper delivery. He reduced the delivery time for a letter from Boston to Philadelphia from six to three weeks and after four years the Post Office made a profit for the first time in its history [16].

In 1757 he was elected by the Pennsylvania assembly to represent the colony in London. Pennsylvania had a big problem. The descendants of William Penn owned most of the land. They were the "absolute proprietors of Pennsylvania" and paid no taxes. It took Franklin five years of delicate negotiating in England before the Penns were forced to pay taxes on all the surveyed land [16].

In 1764 he was back in London as Pennsylvania's ambassador to negotiate for a new charter. This time he stayed for 11 years and was asked by Georgia, New Jersey and Massachusetts to represent them as well. In 1775 war threatened and he returned to Philadelphia. On the day of his arrival he was named delegate to the second Continental Congress.

A year later he was in France as one of three commissioners to seek economic and military assistance for the war against England. For this he was the ideal man. His fame in France was even greater than that of Newton. Nollet, who had published two books against Franklin, was finally forced to give up. One by one the members of the Academy of Sciences had sided with Franklin. He now was the man who could do nothing wrong, there was a veritable Franklin rage afoot in France. His picture could be seen in every house of distinction; stores sold powder boxes with his image on them. The 70-year-old Franklin was the darling of French society [7].

His enormous public esteem helped to win support for America; France signed the treaties in 1778. Franklin stayed on in this pleasant environment, but he was getting old and ill. In 1785, at age 79, he returned, with a large, painful stone in his bladder; he had to be carried to his ship.

He recovered enough to be elected President (Governor) of Pennsylvania and to attend the Constitutional Convention in 1787. Then he returned to Philadelphia and to bed. For the last three years the pain was so bad that he had to be kept on opium. He died in 1790 at age 84.

After Franklin ceased his activities as a scientist his main invention, the lightning rod, was the subject of heated controversy everywhere it appeared. Philadelphia accepted it quickly (Franklin himself was the first one to protect his house). Boston followed and outdid Philadelphia. But soon there was a reaction from the Puritan clergy. Don't interfere with the will of God, they thundered from the pulpit, lightning is a divine punishment. Massachusetts was hit by a series of earthquakes in 1755, which was immediately ascribed to the installation of lightning rods. The pastor of the Old South Church condemned the diabolical invention and observed that there was no escaping God's wrath [18].

The fear arising from religious principles was mixed with more practical considerations in Europe. Franklin's pointed lightning rod actually attracted lightning discharges. The first house of a village to have a lightning rod offered itself to lightning and the people weren't all that convinced that lightning would be safely conducted into the ground.

There was some confusion, partly caused by Franklin himself. In the same letter he had proposed a (grounded) lightning rod and an (ungrounded) rod in the sentry box for experimentation. Franklin and the people at Marly had no idea just how dangerous such an ungrounded rod really was. When a professor at St. Petersburg repeated the experiment, a lightning bolt shot out of the rod right into his forehead and killed him instantly [10].

This accident caused confusion in people's minds. To make matters worse, some of the electricians suggested that the lightning rod should not be pointed but should have a ball at its end so that it would not attract lightning but only conduct it to the ground when it struck. One brilliant scientist suggested that this should be an insulated ball so that lightning would be turned away. The debate went on for years and preoccupied the minds of a great many important people [10].

As we now know there never was any point to the debate. From the distance of a cloud it makes no significant difference if the lightning rod is pointed or round at the top, they both appear very small. Lightning simply travels the shortest distance, it strikes the highest conductor connected to the ground. In the villages and towns of 18th-century Europe the highest conductor was invariably the steeple of the church, a fact that was beginning to become embarrassing to the clergy. The wrath of God was primarily destroying churches.

Worse yet, there was a custom of ringing the church bells to ward off storms. The hapless bell ringer was high up in the church tower, close to the metallic bells. In 1786 a report came out which stated that over the previous 33 years, in 386 recorded lightning strokes, 103 bell ringers had been killed [15].

To make the disaster complete there was also the habit of storing gunpowder in the vaults of the churches. A church being a sacred place, its vaults were considered the safest place to store munitions for the armies. Many a town learned otherwise. In the church of St. Nazaire at Brescia in northern Italy nearly 100,000 kilograms of gunpowder were stored in 1769.

The clergy had refused to have a lightning rod installed. Lightning struck, the gunpowder exploded and destroyed the church and 190 houses around it, killing 3000 people. After the disaster the people of Brescia asked the Royal Society for advice. Franklin was appointed a member of a committee and lightning rods were installed [18].

Because of the emotional and religious resistance, it took more than 50 years for the lightning rod to become commonplace in Europe. But in some places the people also overreacted; everything was to be grounded by Franklin's rod. In 1778 the latest Paris fashion was the "chapeau paratonerre", a ladies hat with a lightning rod and a trailing chain. A little later the inventive French designers offered a similar umbrella for men [15].

Today it is extremely rare to hear of a house (or a church) being damaged by lightning. Considering that there are 16 million thunder-storms on the surface of the earth per year, resulting in some 100 lightning flashes every second [15], this is an enormous tribute to the printer from Philadelphia in America, who had only two years of schooling.

Much Ado About *Almost* Nothing

3. Nine Lives, Nine Discoveries

A French army engineer, three professors, a Danish pharmacist, an emotional French mathematician, a German schoolteacher, an English scientist without formal education, and the odd genius who discovered most of it earlier but wouldn't tell anyone.

Charles Augustin de Coulomb was a French army engineer. He had studied mathematics and civil engineering in Paris and was assigned to the island of Martinique, rebuilding the fortifications destroyed by the English. After nine grueling years he returned to Paris and, still working as an army engineer, started to write some papers in his spare time, mostly concerned with earthworks, dikes and masonry piers and mathematical problems [13].

Charles Coulomb

In 1777 there was a competition, sponsored by the Academy of Sciences, to measure and explain the daily variations in the magnetic field of the earth. The problem was that the ordinary mariner's compass was too coarse to measure the small variations accurately; the pivot of the needle produced too much friction. Coulomb thought of a better way to design a compass. He suspended the magnetic needle on a silk thread and proved that the counter force caused in the silk thread by the rotation of the needle was negligibly small [17].

Coulomb's compass was placed in the Paris observatory, but it was actually of little use: it was simply too sensitive. The needle would jerk when someone would walk up to it, or touch a nearby metal object. Coulomb turned to the problem in 1782. First he tried to eliminate drafts and vibrations, but the needle still moved erratically. He concluded that the problem was due to electrical charge; he needed to replace the silk thread with something that conducted electricity, thus grounding it.

He measured the torsion force of thin piano wire, the force necessary to rotate one end while the other is held fixed. He found two surprising facts: the torsion force was quite small and it was strictly proportional to the angle. For example, if one gram were required to twist the wire by 90 degrees, 2 grams would twist it by exactly 180 degrees [13].

He had set out to produce a pivot with a negligible amount of friction. Now, with his detailed, careful work, he had found a way to accurately measure very small amounts of force. At first he used his torsion balance to measure extremely small mechanical quantities. But then it occurred to him that his balance might be sensitive enough to measure the force of attraction and repulsion of electricity.

In **1784** he built a special torsion balance. Suspended on a thin piano wire was an arm, which carried an insulated ball at one end and a counterweight at the other. The wire was hanging on a knob, which he could rotate and from which he could read a 360-degree scale [17].

Near the ball at the arm was a second, stationary one. He charged both of them with the same polarity and the balls moved apart. Now he turned the knob at the suspension point of the wire and determined the angle required to force the two balls closer together. Having previously determined the force it took to twist the wire, he was thus able to measure the minute force electricity exerted.

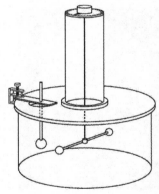

That in itself would have been a substantial contribution. But he found something else. He noticed that, to bring the balls closer together, he needed to twist the knob much more. His mathematical mind began to work and he saw the relationship: the electric force is inversely proportional to the square of the distance. In other words, an electric force behaved in exactly the same way as the force of gravitation, Newton's most famous discovery 100 years earlier. When he modified his torsion balance to measure the force between magnets he found the same law again [17]. Nature had given up one more of its secrets: gravitation, electricity and magnetism all followed the same laws; were they related in some as yet unknown way?

Coulomb measured the force of electricity with a fine piano wire.

Lieutenant Colonel Coulomb received no great recognition for his achievements during his lifetime. He became superintendent of the fountains of France and then Curator of Plans and Relief Maps of the Military Staff. Of his personal life we know virtually nothing; he married late in life, a woman 30 years younger and had to flee Paris during the French Revolution. He died in 1806. Strangely, he never believed in Franklin's theory of electricity [13].

Aloysio Luigi Galvani was born in 1737 in Bologna, Italy. He became a medical student at the University of Bologna, graduated with an M.D. and became a lecturer [9]. After a long time he was promoted to adjunct professor. He was rather shy and introverted, which made him a poor lecturer. But he did some research in anatomy and at least his demonstrations in the lecture hall were interesting. He published very little. This was fortunate for his colleagues, because his writing style was awful. He used long, confusing sentences which often began with one topic and ended with quite another.

But as a researcher in anatomy Galvani became an expert. He was the type of man who could spend hours dissecting delicate organs and enter his observations in his notebook in pedantic detail. In 1780 he was investigating, among other things, the effect of electrostatic machines on nerves and muscles. This had become a popular research subject in medical circles; what intrigued the doctors were the contractions of muscles, which were observed when either a person or an animal was shocked. They reasoned that, by following this lead, they could find out how the nerves and muscles worked together.

The unlucky Luigi Galvani

As it happened one of Galvani's assistants was experimenting with an electrical friction generator as Galvani was dissecting a frog nearby. When Galvani touched one of the main nerves running down the spine, the legs of the dead frog twitched. Galvani was puzzled. He touched the nerve again. Nothing. Someone - by some accounts it was Galvani's wife - pointed out that, at the very moment the frog-legs moved, the electrical machine had generated a spark. Galvani had the assistant turn the crank and touched the nerve with his knife again. The frog-legs twitched. He held his knife by the bone handle. The legs were still. He repeated the experiment over and over again until there was no doubt: the electrical machine somehow caused the legs of the dead frog to move, even though there was no connection. It had been vaguely suspected for some time that the nerves conducted electricity, there was nothing new in this. What was new and startling to Galvani was that the frog-legs moved without actually being connected to the electrical machine [11].

Galvani knew very little about electricity and realized he had to move cautiously. In fact, Galvani was so cautious that he investigated the phenomenon for eleven years before writing a paper about it. He dissected other animals and found that they all responded to electricity for some time after they were killed. Would they respond to atmospheric electricity? He hung skinned frog-legs and limbs of other animals by their nerves with brass hooks on an iron trellis in his garden and grounded the muscles with a wire. Sure enough, during storms they twitched [11].

Then he observed that sometimes frog-legs twitched even when there was no storm. After a while he realized that it most often happened when he pressed the brass hook into the nerve while at the same time the muscle touched the iron trellis. He moved the experiment inside and found that he could produce the same result with frog-legs resting on an iron plate. If the brass hook made contact with both the nerve and the iron plate, the frog-legs moved [11]. His discovery now took on a whole new aspect. No longer was there any external electricity involved. The electrical machine or atmospheric electricity was not needed; the animal limbs seemed to supply their own electricity. "Animal electricity", Galvani called it. This must mean, he conjectured, that it was electricity which made animals - and man - move. It was an awesome thought; if it was true he had found nothing less than the force of life.

He imagined the muscle and the nerve to be like a Leyden jar, with the muscle being the outer conductor and the nerve the inner one. This fleshy Leyden jar stored up energy and held it for some time after death. When the nerve and muscle were connected together, the Leyden jar discharged and the limbs moved [12].

To make absolutely sure he tried other materials, hundreds of them. He found that only metals worked. In fact, two different metals were needed; a single metal did not work. This fact puzzled him but he decided to ignore it. In **1791** he finally found enough courage to write a paper about his research [11]. The paper was so cautiously written and - true to his style - in such cumbersome Latin, that it had a high chance of being buried forever in one of the many volumes of the university research reports. Fortunately - or really unfortunately for Galvani - a few separately-bound copies were printed which he sent to some of the luminaries in the fields of anatomy and electricity. One of them went to Dr. Volta of the University of Pavia, 150 kilometers to the north.

Alessandro Guiseppe Antonio Anastasio Volta was eight years younger than Galvani. He had grown up in Como in the northernmost part of Italy. His family was of the lesser nobility and had almost died out because so many of the Volta boys became priests. Alessandro was the youngest of five boys and was at first considered retarded. He did not speak until he was four, but then his mind suddenly blossomed. By the time he went to school he was among the brightest.

Volta loved drama, had a talent for languages and developed an absorbing interest in chemistry and physics. He studied science and finished with a Ph.D. His first job was at Como as principal of a state school, where he began to experiment in chemistry and physics [8]. He corresponded with as many important people as would answer his letters. With state grants he traveled through Germany, France and England, seeking the acquaintance of the educated, famous and powerful.

After he had become an expert electrician and made some solid experimental contributions, Volta was named professor of physics at the University of Pavia. It was to this well-known, accomplished professor of physics that Galvani sent his paper.

The two men were exact opposites in almost all respects. Galvani was shy, introverted. Volta loved publicity, met people easily and spoke six languages. Galvani was a poor teacher, was afraid to publish and wrote badly. Volta was a popular, effective lecturer, a prolific writer who wrote straight to the point.

Volta at first did not want to read the pamphlet by the professor of anatomy; medical doctors had a reputation for being a bit less than scientific in their writings. And besides, it was written in bad Latin. But his colleagues from the department of medicine at the university urged him to consider it. They felt that someone with expertise in electricity needed to study it. So Volta reluctantly read the paper and, once he got over the bad Latin, found it fascinating. He decided to duplicate the experiments. At first, all went smoothly. Everything Galvani had observed, Volta was able to confirm. But when Volta came to the fact that two different metals were necessary, he hesitated. To him, the electrician, that was significant; unlike Galvani he was not ready to toss this fact aside as a mere nuisance. His training told him to examine it in more detail [8].

He expanded Galvani's experiments to determine the effect of two different metals on living muscles and nerves and found that two coins, one copper and one silver, held together at one end and touching his tongue at two different places produced an unpleasant sensation of taste. He tried different combinations of metals; some produced strong sensations, others weak ones.

But to produce any effect at all there always needed to be two different metals [9].

Volta explored the other senses. He had no luck with hearing but when he took a silver spoon in his mouth and pressed a strip of tin foil both to the spoon and an eyeball, he experienced the sensation of a flash of light.

There was something else that disturbed him about Galvani's theory. In his experiments, the two metals touched a nerve at one end and a muscle at the other. Yet Volta was able to produce twitching by touching the ends of the two metals to muscle tissue or to two nerves only. Galvani's Leyden jar theory was beginning to fall apart. Volta suspected that the animal limbs were only reacting, that the source of electricity was not in the animal, but in the two metals.

Alessandro Volta, who corrected Galvani's finding and thereby ruined his career.

Volta wrote a paper in **1794**, "Account of some discoveries made by Mr. Galvani, with experiments and observations on them" [38]. As he was already well known and respected, he got it published in the most prestigious of all scientific journals, the Philosophical Transactions of the Royal Society in London. He was openly critical of Galvani's theory.

Galvani was no fighter and did not want to respond. But his nephew, Giovanni Aldini, who later became professor of physics at Bologna, was more than willing to take on the fight for his uncle. There was also an old rivalry between the universities of Bologna and Pavia. Aldini formed a society at Bologna whose express purpose was to attack the infamous professor of physics at Pavia [9].

In the meantime something curious had been found at Bologna: sometimes, under the right circumstances, dead frog-legs could be made to twitch when the nerve of one was held to the muscle of another, thus completing an electric circuit without the use of any metal at all, clear proof that animal electricity did exist [12]. A paper describing this experiment was published (anonymously) and Aldini wrote pamphlet after pamphlet attacking Volta in flowery prose.

The controversy engulfed the scientific circles all over Europe. Debates were held in journals and at scientific gatherings. There was not enough evidence for either case, but more than enough for tempers to flare.

The discovery of animal electricity without metals was eclipsed by

Volta's subsequent announcements. Within a year he was able to refine his instruments and prove that a junction of two different metals indeed produced electricity. Although Volta still - and always - gallantly called it the galvanic effect, the term "animal electricity" slowly began to disappear and was replaced by "metallic electricity" [8].

Galvani's fortune declined. He had lost his wife, childless, in 1790. The scientific world was abandoning him. In 1796 Napoleon invaded Italy. Bologna had been part of a church-state; its authority was the Church of Rome. Galvani was forced to swear allegiance to the new Cisalpine Republic. He refused; being an officer of the university he had already sworn allegiance to Rome. As a consequence he lost his job and, stripped of his income, he moved into his brother's house. Napoleon went campaigning in Africa and, in the bureaucratic confusion that followed, Galvani's superiors were able to restore him to his post in 1798. But it was too late. A few months later he died, heartbroken, bitter and penniless [9].

Volta, the man of the world, had few problems with the changing political situation. When Napoleon invaded, he was a member of a delegation to honor him and became an official in the new government. When the Austrians recaptured Pavia and held it for 13 months he had no trouble moving with the tide again.

Despite the political upheaval, Volta made progress on his research. He had found that a moistened piece of cardboard between the two metals greatly enhanced their effectiveness. To the list of metals that produced electricity he added graphite and charcoal and he ranked all materials in effectiveness. The further apart on his list the materials were, the more electricity was produced. Yet even the best of these combinations of materials was not very useful. Their effect could be felt on the tongue, but it did not have the shocking power of a Leyden jar. So Volta stacked metal discs with moistened cardboard on top of each other, the first embodiment of the modern battery. With 30 to 40 of these

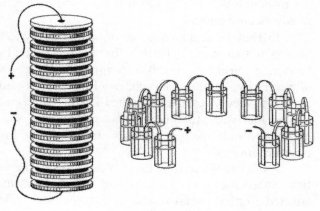

Volta's batteries.

combinations in series he could now feel the "tension" with his hands [8].

There was one remaining problem with Volta's invention. The cardboard discs soon dried out and the electricity disappeared. To counter this he put plates of two different materials in glass cups and added salt water. If there was any tendency for the liquid to dry out, he could simply refill the cups.

What had Volta invented? Before his battery appeared, electricity was produced by rotating a globe or disc and could be stored in a Leyden jar. After a spark was drawn, or the Leyden jar discharged with a wire or a finger, the electricity was gone. With Volta's battery the electricity remained; it could be used many times, yet the electric "tension," as Volta called it (and for which we now use the term voltage in his honor) would remain strong. This difference was so startling to Volta that he was convinced he had found perpetual motion. He was, of course, a little premature in his conclusion. As millions of users of transistor radios, flashlights and toys know, batteries do wear out.

But Volta's battery had another and far more important feature: it produced a steady, continuous current. The term "current" was not known then in electricity and few people had any notion of what it was or grasped the idea that there was something else beside "tension". But the evidence of electric current was there. One needed only to connect a thin wire to the battery: it would heat up and stay heated as long as the battery had any charge in it.

Current had been flowing from the Leyden jar too, but it was of such short duration that it could never be observed. What Volta had in fact pried out of nature was not so much the electric battery as the far more important steady electric current.

In **1800** he sent an account of his discovery to the president of the Royal Society in two parts [39]. Even before the second part arrived, Volta's paper had become a sensation. His battery immediately transformed the entire science of electricity. Within a month, two Englishmen made a major discovery with it. When they stuck the two wires from a battery into water, gas bubbled up around the wires; the current was splitting the water into its two elements, hydrogen and oxygen. This triggered a torrent of discoveries in electrochemistry [8].

As Galvani's star declined, Volta's rose. The year after he published his first observations on the ill-fated Galvani discovery, Volta received the coveted Copley medal of the Royal Society. After his publication on the battery, Napoleon asked him to come to Paris and give a demonstration to the National Institute of France, the successor to the Academy of Science.

Napoleon personally assisted him and had a gold medal struck in his honor. The First Consul (and later emperor) was so enamored by Volta that one day, as he was walking up the steps to the library of the National Institute and saw an inscription "To the Great Voltaire", he ordered the last three letters chiseled out. In 1806 Napoleon, who had crowned himself King of Italy, made Volta a Knight of the Iron Crown. In 1809 Napoleon appointed him a Senator (he never spoke in the senate) and in 1810 he became Count Volta. After this he had the income of a rich man. Sadly, as soon as he became a celebrity his scientific work stopped [8].

Not once during the entire animal vs. metallic electricity controversy and for another ten years after Volta's death in 1827 did it occur to anyone that the two men might have observed two different effects. Galvani had chosen to ignore the inconvenient fact that the presence of two different metals was important. Volta concentrated entirely on the effect of the metals and ignored Galvani's later discovery that the same twitching effect could be produced without any metals.

Galvani had stumbled across two effects simultaneously. Volta isolated one of them, the electricity produced by two different metals. As it turned out, Volta's explanation was wrong; the effect only works in the presence of an electrolyte (e.g. in the moistened cardboard disk), being primarily a chemical reaction.

Galvani's explanation was, of course, wrong too, but he came much closer to the truth than anyone suspected. There is indeed such a thing as animal electricity, though it is not the "force of life". Muscles, especially when injured, build up small electric potentials (about 70 millivolts). Nerves are electrical conductors and have a trigger potential that is slightly smaller (about 30 millivolts). If a dissected muscle touches a nerve, it can trigger it with its own potential [12].

What Galvani had discovered, though he did not know it, was the electrical activity of muscles, a discovery that ultimately led to the electrocardiogram and the electroencephalogram. Judging from the distance of 200 years, Galvani's work was as important as that of Volta. Yet Galvani reaped nothing but heartache, while Volta was showered with honors.

Hans Christian Oersted (or Ørsted) grew up on the island of Langeland in Denmark. There was no school, so he learned reading and writing in Danish from the wife of a wig maker, arithmetic and German from the wig maker himself, geometry and drawing from the town surveyor and French and English from the town mayor. At age 17, having read every book available locally, he passed the entrance examinations for the University of Copenhagen. At the university he supplemented a small government stipend with tutoring. He received his Ph.D. in pharmacy in 1799 [10].

For a year he managed a pharmacy in Copenhagen, while he began to experiment with electricity. Then Oersted wandered through Germany and France, absorbing the science practiced there. At heart Oersted was a philosopher; his Ph.D. thesis had been about Kant's philosophy, with a heavy dose of metaphysics. During his travels he was strongly influenced by the theories of a Hungarian chemist, Jakob Joseph Winterl. Winterl had devised a fantastic system of chemistry, which was entirely based on fantasy, not facts. Oersted returned to Copenhagen an enthusiastic supporter of Winterl and made the mistake of writing a paper on this subject. When he applied for the position of a professor he was rejected; the faculty found him too gullible [10].

Hans Christian Oersted.

But Oersted matured. He became disenchanted with Winterl's theory and, when he reapplied in 1806, was made professor of physics. He now lectured and did research in electricity and found a subject, which appealed to him strongly: the relationship of electricity and magnetism [31].

Was there any relationship between the two phenomena? Researchers had attempted to find one for well over fifty years but had drawn a complete blank. They had tried to deflect compass needles by holding them to batteries, wires, spark generators and everything that was remotely connected with electricity, but had never been able to discern any effect. Yet it was known that, sometimes, pieces of iron struck by lightning turned into magnets, or lightning could reverse the polarity of a compass. Franklin had magnetized a needle by discharging a Leyden jar through it. Papers were written on the subject, mostly vague and mostly wrong. Oersted, too, wrote a paper on the subject (in 1813). He expounded the theory that, since a thick wire connected to a battery produced heat and a thinner wire produced light (i.e. a flash of light when it burned up), a still thinner wire would produce magnetism. Poor Oersted was on the wrong track.

Although cured of Winterl's theories, Oersted was in fact still under the influence of the German "Naturphilosophie", which held that the universe was a divine work of art, governed by a few simple forces. There was no doubt in Oersted's mind that heat, light, electricity and magnetism were only different forms of the same power.

Oersted and his contemporaries now had a significant advantage: Volta's battery. The Leyden jar had produced a significant voltage, but the discharge happened so quickly that nothing other than the effect on a human body could be observed. Volta's battery, although much smaller in voltage, provided electricity which lasted much longer.

In April of **1820**, during a lecture, Oersted made a discovery. He was demonstrating the heating of a wire connected to a battery when he saw the needle of a compass nearby move slightly. At first the observation seems to have made little impression on him, he did not pursue it immediately. But in July he looked at the phenomenon again. Using a more powerful battery he now could clearly see the reaction of the compass needle. There were two reasons why the phenomenon had been missed so far. First, it was not the tension (voltage) which caused magnetism, but current flowing through the wire. Oersted

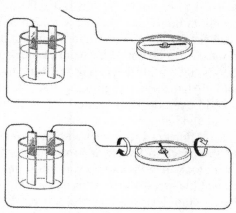

Oersted discovered that an electric current produces a magnetic field. But it was not *along* the wire as everyone expected but *around* it.

had only a vague notion of current (he called it electric conflict), but he correctly surmised that whatever was flowing through the wire caused the deflection. Second, the compass needle was deflected *perpendicular* to the wire; the magnetism was around the wire, not along it. This was totally unexpected; in the world of mechanical physics there had never been a force that acted perpendicular to another force [32].

Oersted investigated the relationship as thoroughly as he could. He tried various materials for the wire and found they all worked the same way. He reversed the current in the wire and found that the compass needle reversed direction also. He verified that the magnetism created by the wire penetrated wood, glass, marble, clay, water etc., suggesting that the magnetism produced by the wire was the same kind as that produced by a magnet. He wrote a brief paper, which struck the electrical community like thunder.

The French physicist Francois Arago ran across a copy of Oersted's paper in Geneva and brought the news to Paris. There, at the Academy of Sciences, he demonstrated the experiment in September **1820**. In the audience sat a 45-year-old mathematician named Andre Marie Ampere.

Ampere was born in Lyons. His father was a rich merchant who soon after the son's birth retired from business and moved his family to a small village outside Lyons. Ampere was a puny, sickly, precocious child; he learned to calculate by himself with pebbles before he could talk [2]. He was taught at home and became a wizard in mathematics. At age 14 he wanted to read the works of Euler and Bernoulli and when told by the librarian that the books were in Latin, he went home, studied Latin and was able to read the books one month later. By this time he had read the 20 volumes of Diderot's and d'Alambert's freethinking Encyclopedie cover-to-cover [37]. He grew up in a rigid catholic family; the reading of the Encyclopedie caused doubts in his mind and his problems started.

After the French revolution broke out Ampere's father became a justice of the peace in Lyons, a stronghold of the royalists. In 1793 a revolutionary army took the city and he was guillotined, for no other reason than that he had issued the arrest warrant for a revolutionist. Ampere withdrew into himself, he had a complete mental breakdown, not speaking or doing anything for more than a year. He finally came out of it when he developed an interest in botany [21].

He was extremely nearsighted and for some reason was not fitted with glasses until he was 18 years old. During his youth he could see nothing beyond a few meters and would hold books very close to his eyes [37].

André Marie Ampere, a brilliant but very emotional mathematician.

At 21 he fell in love and married. It was the only happy time in Ampere's life. Alas, the happiness didn't last long, his wife died four years later. He married again, but his second marriage was a disaster from the start; within a few months he was divorced. Ampere moved to Paris where he became an assistant professor in mathematics and chemistry at the Ecole Polytechnique.

To his students Ampere was a ridiculous figure, wearing ill-fitting suits stained by chemicals. He would pull out his handkerchief and then absentmindedly wipe his brow with the blackboard eraser and clean the blackboard with the handkerchief. Once, it is told, he was walking as he was struck by an idea to solve a mathematical equation. He pulled out a piece of chalk and

　　　　Much Ado About *Almost* Nothing

started writing the equation on the back of a carriage. Just as he was coming to the solution, the carriage pulled away.

He was pathetically emotional, especially when it concerned the miseries of mankind. One day, after the news of a defeat of Napoleon's army had reached Paris, he began his lecture on chemistry in these words: I fear I shall fail today in clearness and method; I have scarcely strength to collect and connect two ideas; but you will pardon me when you learn the Prussian cavalry were passing and re-passing over my body all night" [2].

But he was an exceptional mathematician. When he learned of Oersted's experiment his mathematical mind began to grasp relationships. He had not done any work in electricity up to this point, but the relationship between electricity and magnetism suddenly interested him. He went to his laboratory and first improved Oersted's experiment, using a second compass needle fixed to the first one some distance apart and pointing in the opposite direction, thus eliminating the influence the effect of the magnetic field of the earth. Now the needles showed an exact 90-degree position to the wire when the current was turned on. Next he realized that such a needle could not only be used to indicate when there was a current flowing, but to measure the magnitude of it [40].

Being able to correctly map and thus understand the magnetic effect around a wire, Ampere now made his most practical discovery: if he wound the wire into a helix (or coil) the magnetic fields of the turns would all point in the same direction and thus add. In this way the magnetic effect *inside the coil* was multiplied by the number of turns [40].

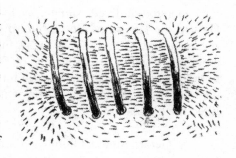

Ampere's coil, which multiplies the magnetic field inside by the number of turns.

He presented papers at the next few meetings of the Academy, starting just a week after he first learned about the subject [20]. Oersted had not provided any mathematical underpinnings for his discovery; Ampere did. In the incredibly short time of three weeks he deduced almost the entire mathematical relationship between electricity and magnetism. He demonstrated that a carefully suspended wire, connected to a battery, would orient itself to the magnetic field of the earth [42].

Having seen a wire behave like a magnet, he came to the conclusion that electricity was the fundamental cause of all magnetism. In his concept a piece of steel, which had become a permanent magnet had electrical currents flowing inside it all the time. Even the magnetism of the earth, in his view, was caused by electrical current flowing in the center of the earth. As always, his theory was supported by elegant mathematics [42].

He went on doing research in electricity for another seven years. But he was not a steady worker. All his life, Ampere's work was accomplished in brief, brilliant flashes of hectic activity, followed by periods of apathy. As he grew older, the periods of apathy grew longer; he lost interest in a subject just as he was becoming eminent in it. He would then buy books but leave them unopened on the shelf [2]. One of his many unfinished projects was a classification of all sciences into 128 categories, a job that, by the 19th century had become far too large for one man.

More and more he was assailed by religious doubts and if there was anything Ampere couldn't stand it was doubt. "Doubt", he said, "is the greatest torture man can endure on earth" [21]. As he grew older he became increasingly irritable, undiplomatic and had periods of rage. His face became distorted and his voice changed. He died in 1836, at age 61. He chose his own epitaph: "Happy at last" [2].

Oersted outlived Ampere by 15 years. His was a happy, productive life. He became Knight of the Legion of Honor, Knight of the Prussian Order and received the Grand Cross of Danneborg. His 50th anniversary of his association with the University of Copenhagen was celebrated as a national holiday and the government gave him a country house near Copenhagen. He died in 1851; behind the coffin walked 2000 dignitaries as 200,000 people lined the procession [10].

Georg Simon Ohm was born in 1789, the son of a locksmith. His father, a self-sacrificing, self-taught man with a strong interest in science, educated his children in mathematics, physics and chemistry. Even though Ohm went to public school, he received most of his education at home [30].

At age 16 he enrolled at the University of Erlangen. But he was not yet ready for serious study; after three semesters the displeased father took the boy out of the university because the latter had spent too much time dancing, ice skating and playing billiards. Ohm left home and went to Switzerland where he taught mathematics for two years at a private school. After this stint he decided that he wanted to study after all, went back to Erlangen and received his Ph.D. in 1811. His ambition now was to become a professor at a university.

He started as a lecturer at the university, but the pay was so poor that he couldn't afford to continue. He switched to a low-prestige school, which offered better pay but closed after three years. In 1817 he found a job as professor at a middle school in Cologne. He had the title, but it was not a university [29].

Ohm was an outstanding, inspiring teacher. He had the knack to make even a usually dull subject such as geometry attractive to his students. But he was stuck in the level below the university. He knew that to move up he needed to do research. Even in those days professors needed to publish to be promoted. Simply being a good teacher, unfortunately, never accounted for much.

Georg Simon Ohm found the fundamental law of electricity, but his writing was so obscure that it was dismissed as unimportant.

Luckily, the school in Cologne had a well-equipped laboratory. Ohm decided to investigate the behavior of electricity. He didn't know very much about it, but he believed that he could apply his main strength, mathematics, to advantage [44].

So he began to connect wires of various lengths and different materials to batteries. He had constructed an instrument similar to Coulomb's torsion balance, a magnetic needle suspended on a thin wire. With the needle close to the wire he could measure the angle of deflection and thus get an indication of the strength of current [30].

But the experiments didn't work out too well. There was no place yet to buy a battery; an experimenter had to build his own, a messy affair with glass-cups and acid. The ones that Ohm made were too poor to get accurate results. The strength of the current would gradually drop as he was taking the measurement. Ohm tried to make the best of an unsatisfactory situation. He applied ingenious correction factors and found a relationship between the material, the wire size and length, the voltage and the current, but his conclusions were complicated and badly flawed. He wrote two papers about his experiments and calculations and they were published in **1825** in a scientific periodical [29].

One of his readers suggested that he replace the battery with a thermocouple. Four years earlier the Estonian Thomas Seebeck had discovered that some combinations of two metals produced a small voltage if the two metals were at different temperatures. This thermoelectric effect had the advantage that the voltage remained more stable after the wire was connected.

Ohm tried it and it was indeed exactly what he needed. He placed one junction of the metals in ice, the other in boiling water and could, for the first time, measure the current reliably. He moved in with his brother in Berlin and took a year's leave of absence at half pay to finish his investigation and write a book about it [26].

Ohm had found that the current in a wire was the same at any point; many people had assumed the current would drop in magnitude as it flowed toward the negative terminal and became exhausted (akin to the Colorado river, which never reaches the ocean but ends as a trickle in the desert). More importantly, the current he measured was directly dependent on the amount of voltage.

Picture a lake high up in the mountains and a river flowing from it to the ocean. The altitude of the lake is the voltage (or the potential). The flowing water in the river is the current. If the riverbed is wide and smooth, the water finds it easy to flow downhill; if the bed is narrow and rocky, the water encounters resistance and the flow is reduced. In electrical terms, if you have a small voltage and a thin wire that conducts poorly, you only get a small current. If you want to increase this current you either need a higher voltage or a thicker wire of better-suited material, or both. This is nothing less than the fundamental electric law:

$$\text{Current} = \text{Voltage} / \text{Resistance}$$

But you won't find this equation (now known as Ohm's law) in Ohm's book; it is buried deep in complicated mathematics. He talks around it, never quite getting to the point. Worse, his two previous papers, which were slightly wrong, had received the wide distribution of a major periodical, while his book sold poorly. For this reason most electricians knew only his preliminary work [29]. To make matters worse, a well-known professor of physics, who didn't think much of Ohm and less of his book, attacked him.

Ohm was now 38 years old and had expected to receive an offer for a professorship at a university after the publication of his book. No offer came. Dejected, he resigned his job in Cologne and stayed in Berlin, tutoring boys with little ability and less desire [29].

But slowly the acceptance of Ohm's work changed. A Frenchman named Pouillet verified his work in clearer terms, which caused Ohm's detractors in Germany to pay attention. Six years after the publication of his book Ohm was offered a position at the Polytechnic School of Nuremberg. Eight years later he received the Copley medal from the Royal Society in England. In 1852, when he was 63 years old, he became a professor at the

University of Munich. He had finally achieved his life's ambition - two years before he died [44].

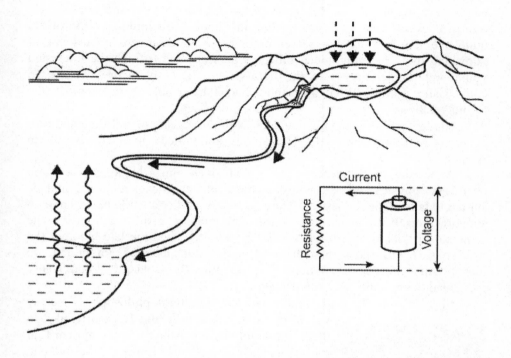

Ohm discovered (but was unable to describe) that electric current is like the flow of water. Current increases with the altitude of the reservoir (the voltage of the battery) and by making the river wide and smooth (decreasing resistance).
There needs to be a return path in either case.

Joseph Henry was born in Albany, New York, in 1797. Henry's father was an alcoholic who worked only sporadically; when Henry was seven, he was sent to an uncle's house. At 10 he worked mornings in a general store and attended some classes in the afternoon, receiving only a rudimentary education. The boy was a romantic daydreamer; he disliked the standard, stifling grade-school education intensely [6].

But the story takes a strange turn. It seems he had a pet rabbit that one day ran underneath the church. Henry crawled after the rabbit and was attracted by light shining through the cracks of some loose floorboards. Lifting them up he found himself inside the one-room village library. He began to explore the shelves and started reading one of the books. He returned several times to finish reading this book and continued on to others until one day he and his secret entrance were discovered. Instead of punishing him, the enlightened village elders decided to give the interested young lad regular access to the library. In this way Henry became an enthusiastic reader. What the school hadn't been able to accomplish, the forbidden library had [6].

His father had died and, at age 14, with three years of full-time and four years of part-time schooling, Henry was apprenticed to a silversmith in Albany.

A boarder in his mother's house had given him a beginner's book on chemistry, physics and astronomy. These subjects intrigued Henry and he started to take night classes at Albany Academy. After a while he was able to support himself by teaching English grammar at district schools. The principal of Albany Academy, Dr. Romeyn Beck, liked the late bloomer (he was considerably older than the rest of the students) and got him a job tutoring the children of General Stephen Van Rensselaer. When Henry graduated, he became Beck's assistant [6].

In 1826 the chair of "mathematics and natural philosophy" became

vacant at Albany Academy and Beck offered it to the 27-year-old Henry. Albany Academy had four professors and 150 students. Henry's duties were to teach mathematics, physics and chemistry. Before the school year began he traveled to New York and there he saw a model of a magnet invented by William Sturgeon, who had found that steel needles in the core of a coil enhanced the magnetic effect. Henry decided he wanted to duplicate this interesting device and investigate its principle further. But there was no space for a laboratory in the small building of Albany Academy. Even his lecture room was occupied from six in the morning to late at night; the only time he could convert the lecture room into a laboratory was during school vacations in August [6].

Joseph Henry

Henry began his research on the electromagnet in August 1826. The fad of the time was to build huge batteries, occupying entire basements, to obtain as much power as possible; Sturgeon's magnet worked well only when it was connected to such a powerful battery. Henry had neither the money nor the space to construct a large battery, thus he decided to investigate if he could increase the efficiency of the electromagnet by understanding how it worked [19].

Sturgeon had insulated the iron core with varnish and then wound bare copper wire over it, so that the shape of the winding was exactly a helix, just as Ampere had suggested. Henry, the novice electrician studied the arrangement and came up with a simple idea which had not occurred to any of the brighter lights in the field: if he could insulate the wire it would be possible to wind more than a single layer onto the iron core. So he carefully wrapped the wire with silk ribbon, hundreds of meters of it, and wound the iron core tightly with the insulated wire. His experiments lasted for four summers and he was making remarkable progress, but he felt he was not ready to tell the world. All his life Henry had to be prodded into publishing his research, he never was certain when he had anything new to tell. In addition, his irregular education had left him with a bad deficiency in spelling and grammar, a handicap, which he only gradually eliminated during his lifetime [19].

But after the summer vacation of 1830 he was forced into action when he found a paper in a European periodical describing an electromagnet that could lift 35 kilograms (70 pounds), still using Sturgeon's single layer winding, but powered by a huge battery. Henry had a much better story to tell. With a puny battery he had been able to lift 350 kilograms (700 pounds)! Hastily he wrote a paper and submitted it to the American Journal of Science, published by Professor Silliman of Yale. Besides the more efficient winding structure he reported another, subtler but equally important improvement. He had wound separate layers of wire, each of them with its ends sticking out, so he could connect them individually. He compared the

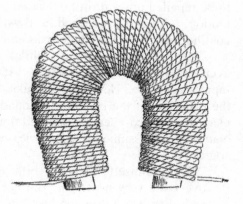

Henry's electro-magnet.

lifting power of the electromagnet when the respective ends of its windings were all connected together (he called this a "quantity circuit", we now know it as parallel connection), and when the windings were connected so that the current would flow in sequence through each of them (which he called an "intensity" circuit, now known as series connection). In this way he was able to match the battery (whose cells could also be connected in either series or parallel) to the load and use the energy in the battery with maximum efficiency. What happened in fact was that Joseph Henry was intuitively beginning to understand Ohm's law, though he hadn't yet heard of Ohm[6].

Professor Silliman was delighted with the paper and commissioned Henry to make a bigger electromagnet, powered from a larger battery. This magnet was able to lift more than 2000 kilograms (4000 pounds). Henry, teaching in an obscure school without a proper laboratory and without a research fund, had outdistanced all other electricians by a large margin [19].

With his stunning achievement of the powerful electromagnet Henry suddenly became famous. He had given the world the first practical electrical device [23].

A few months after his path-finding paper he published a second one in the Journal of American Science. In this paper he described an electric motor. Suspended in the center on a pivot there was an electromagnet with two windings. The ends of it faced two permanent magnets and the wires of the windings made contact with two batteries just before the ends of the electromagnet touched the permanent magnets. The polarity of the battery was such that the permanent and electromagnets repelled each other. Thus, as each arm moved down, its wire made contact with a battery, which caused it to be repelled upward. In this way the arm rocked back and forth in constant motion. Later he attached a flywheel, which caused the bar to rotate continuously, until the battery was exhausted.

Henry called his motor a philosophical toy and he didn't think it would ever amount to anything practical. At that time the steam engine was finding rapid acceptance. He made a calculation, comparing the cost of coal to that of the materials of the battery and found that, to produce the same energy, his electric motor cost several hundred times as much to run than a steam engine. Naturally, he assumed that batteries would remain the only electrical power source [23].

Henry himself was to overthrow that assumption with an awesome discovery the very next year. But because of his tardiness in publishing his research, Faraday in London beat him to the press. As far as we can tell, the two men came to their conclusions within days of each other, and for this reason each deserves credit. To give equal time, we shall now move across the Atlantic Ocean and continue the story of Joseph Henry's career later.

Michael Faraday was six years older than Henry; he was born in 1791 in Yorkshire. His father was a blacksmith; during the depression that followed the French Revolution the family moved to London - near Manchester Square - in the hope of finding employment. But things went badly for Faraday senior, so the family went through a period of dire poverty. Michael also received only a rudimentary education, just barely learning to read, write and calculate [41].

At 12 he was taken out of school to earn his keep, working as an errand boy for a bookbinder and bookseller who also lent out newspapers. It was Michael's job to shuttle the newspapers between the readers. The bookbinder was impressed by the bright and alert teenager, who showed a great deal of interest in the books being bound and sold. At 14 he took him in as an apprentice, waiving the customary fee. It was the ideal environment for a young man hungry for learning.

Faraday read books on chemistry and found the subject fascinating. When he had to bind the volumes of the Encyclopaedia Britannica he studied the section on electricity and decided he wanted to repeat some of the experiments. Saving up his meager allowance for weeks he managed to buy a large glass bottle, which he converted into an electrostatic generator, and a smaller bottle, which became the Leyden jar. He discussed his experiments with his friends at every possible occasion and one of his customers loaned him the money to attend a series of lectures [41].

After the apprenticeship was over he wrote to the president of the Royal Society to seek a job - any job - in science. There was no reply. A few months later he attended some lectures by the famous chemist Humphry Davy at the Royal Institution (a center for lectures and research in London), then wrote out his notes in neat longhand, bound them into an attractive book and sent them to Davy. Davy was kind enough to talk to him, but he didn't have a job for him at that time. Davy also recommended that Faraday should not give up such a steady job as a bookbinder for the insecure world of science [41].

Shortly after Faraday's disappointing encounter with Davy, an assistant was fired at the Royal Institution because of a brawl and Davy remembered the young bookbinder who had taken the trouble to present him with the notes of his lectures. Thus, in 1813 the 21-year-old Faraday got a job as assistant at the Royal Institution at a salary of one guinea a week, two rooms in the attic, plus fuel and candles. Faraday's duties were to keep the place clean and help the professors prepare the lectures [33].

Davy had married a rich widow the previous year. Five months after Faraday started, Davy decided it was time to take an extended trip to the continent and Faraday was invited to come along as part of the entourage.

Luck was shining on Faraday; here was an opportunity to travel through Europe in style and visit all the scientific greats. The party went to Paris, a trip only possible to a scientist of Davy's renown, since France and England were at war at that time. In Paris Faraday met, among other famous people, Ampere. From Paris they went to Italy and Switzerland. Eighteen months after the beginning of the trip they returned to London and found the Royal Institution in one of its frequent financial difficulties. Faraday wasn't sure he would have a job there, but after a month he was offered an even better one: assistant and superintendent of the laboratory [41].

Slowly, over the next few years, Faraday began to do research in chemistry, first as Davy's assistant, then more and more on his own. In 1817 he published his first papers, nothing outstanding, but solid, original research. Considering that he had only a bare minimum of grade-school education and no formal training in the field of chemistry except what he had learned from Davy and read in books, it was a remarkably fast rise.

Michael Faraday

In 1821 he married Sarah Barnard. Like Faraday, Sarah was a kind, modest person and belonged to a religious sect called the Sandemanians, whose members believed in the literal interpretation of the Bible and felt that the Christian faith had been corrupted by the involvement of the state [43]. Faraday was a devout Sandemanian and served as an elder of his assembly. He never seems to have experienced any conflict between his religion and science. He only spoke of his religion if he was drawn into a discussion and even then volunteered little information.

The marriage of Michael and Sarah Faraday was a happy one, even though it didn't produce any children (and Faraday was especially fond of children). The couple moved into two rooms on the top floor of the Royal Institution, and lived there for 37 years.

The year 1821 also marks Faraday's entry into electricity. With Davy he had been involved in electrochemistry, the separation of chemical solutions with electric current, which led to the discovery of a large number of new elements. But until Oersted found the reaction of the compass needle and Ampere explained it, Faraday was exclusively a chemist. Davy brought the news of Oersted's and Ampere's discoveries, but he had misunderstood much of it. He and his friend Wollaston, a scientist well known in the circles of the Royal Society, speculated that a wire carrying a current would rotate around its axis if suspended freely, since the magnetism was circulating around it. They set up an experiment to prove their theory, but the wire refused to rotate. Faraday was not present during the experiment, but he heard the two men discuss it afterwards [41].

A few weeks later Faraday was invited to write an historical account of electromagnetism for the Annals of Philosophy. To get to understand the subject he repeated the major experiments; as a newcomer to the field he examined the theories critically and soon became convinced that no one really understood magnetism.

Ever since Newton, one of the most persistent subjects for debate had been "action at a distance". Did gravity act at a distance or was there some "imponderable fluid" even in vacuum? Franklin's friend Hopkinson had found that he could discharge an electrically charged cannonball with a pointed needle even before the needle came into contact with the ball. Was this action at a distance again? The same question was raised about magnetism. It appeared that the magnetic force penetrated all non-magnetic materials, even vacuum.

One of the experiments Faraday repeated consisted of sprinkling iron powder on a sheet of paper held close to a magnet. In this way the paths of the magnetic force become visible, the iron particles orient themselves in striking patterns between the poles. Similarly, if a hole is punched into a sheet of paper and a wire, carrying a current, is looped through the hole, iron powder on the paper shows a pattern of circles around the wire. These images told Faraday that there must be lines of magnetic force. To Faraday these lines were a physical reality, a force outside the magnet or the wire, penetrating even vacuum [41].

This insight led him to try a somewhat different version of Wollaston's and Davy's experiment, and he found that he could get a small magnet to rotate around a wire, or a wire around a magnet. He published his finding immediately.

The publication was a shock to the scientific community in London. Wollaston had been talking about rotating wires and here was a mere laboratory assistant who had stolen Wollaston's idea and published it. They couldn't see the subtle variation, brought out by an important difference in perception of the fundamental behavior of magnetism. It took Faraday two years of frantic explaining to make his contribution understood and to clear his name [41].

During the next ten years Faraday periodically came back to his initial experiments in electricity. He was still a chemist, busy with such projects as an extensive investigation of the alloys of steel, but he couldn't quite leave electricity alone. He was dissatisfied with Ampere's explanation of electromagnetism. Ampere had supported his conclusion with dazzling mathematics. Faraday, who never used mathematics when he could avoid it, had seen his conception of magnetic lines of force produce tangible results. More and more he became convinced that these magnetic lines were a fundamental phenomenon [1].

More important and titillating was another problem: Oersted had shown that electricity produced magnetism. Could magnetism produce electricity? Faraday pondered about this subject, as did many other researchers. But a magnet held to a wire produced no discernible effect on a galvanometer, no matter how strong the magnet was. Even with a coil, a magnet showed no effect. This was baffling and became somewhat irritating as time went by. Sturgeon and Henry had shown that electric current could produce a large amount of magnetism, but no amount of magnetism could produce even the slightest electric current.

Faraday decided that an electromagnet might be a tool to investigate the fundamental nature of magnetism. He suspected that the iron core might be in a "peculiar" state when it became magnetic [18]. In **1831** he took an iron ring and wound two coils on it; one he connected to the battery, the second to a galvanometer. He fortunately made contact to the galvanometer first. When he connected the battery, he noticed the needle of the galvanometer jump. Surprised, he disconnected the battery and the needle jumped in the opposite direction [15].

At first the erratic behavior of the galvanometer made no sense to Faraday. He repeated the experiment and then suddenly saw what was happening. The second coil produced an electric current whenever there was a *change*. It was not magnetism itself which produced electricity, but a *change in the magnetism*. He took a permanent magnet and moved it through the center of a coil to which a galvanometer was connected. The galvanometer responded to the movement of the magnet, its needle moved to one side when he pushed the magnet in, to the other when he pulled the magnet out. The faster he moved the magnet, the stronger the deflection of the galvanometer [15].

Henry, across the Atlantic, had arrived at exactly the same conclusion in exactly the same way within days of Faraday. The two scientists never quarreled about priority of discovery. But while Henry hesitated, Faraday published his findings immediately, and for this reason he is almost universally acknowledged as the discoverer of electrodynamics. But credit for this achievement should belong to both. They had independently and simultaneously found a system to produce electricity from mechanical energy;

it was the first faint forerunner of today's electric power generators and this discovery of *electromagnetic induction* became of fundamental importance for the growth of electrical science.

The discovery strengthened Faraday's belief in the concept of the magnetic lines of force. He was able to show that electric current was produced whenever a wire (or a coil) moved through, or "cut", a line of magnetic force; the faster these lines were cut, the larger was the resulting current.

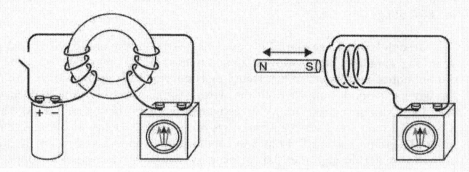

Faraday (and Henry) found that it was a *change* in a magnetic field that produced electricity. The needles twitched when the current was either turned on or off or when the magnet was moved.

Faraday now spent the major part of his time on electromagnetism. He was elevated to professor at the Royal Institution, refined his concept of the magnetic lines of force and expanded it to the forces of electricity and gravity. But in this he was alone. It was not until 1845 when William Thomson (who later became Lord Kelvin) provided the mathematical underpinnings that Faraday's theory of magnetic lines found grudging acceptance. In 1855 it became the basis for Maxwell's great mathematical work on electromagnetic fields. But Maxwell's work was so abstract that Faraday himself was never able to understand it [41].

Faraday managed to add a remarkable number of other discoveries to his already outstanding record. Among other things he proved that all materials are in some ways magnetic and that some materials, when subjected to a magnetic field, rotate polarized light [1].

He was an outstanding lecturer and drew large crowds to the Royal Institution and thus, like Davy before him, helped it survive many a financially precarious situation. He could explain even the most complicated subjects in simple language, inspiring his audience and making them wish they were scientists, too. But his favorite lectures were for children each year at Christmas, when he introduced his young audience with open mouths to the wonders of science. Not having any children of his own, he seems to have lavished special love and preparation on these lectures.

A simple man to the last, he turned down a knighthood and the presidency of the Royal Society. Queen Victoria placed a house at Hampton Court at Faraday's disposal when he was 67 years old and began to suffer from loss of memory [1,16], quite possibly from life-long mercury poisoning. By age 72 he could no longer function, having become quite senile. He died in 1867, at age 75.

Joseph Henry's account of his own discovery, tardily published almost a year after Faraday's paper, showed an additional phenomenon, which Faraday did not notice until later. When Henry disconnected a large enough coil from the battery, a spark jumped from the wires. Sometimes this even happened with long, straight wires. Henry investigated the phenomenon and found that a coil (or a long wire) was capable of **storing** current. Henry also found that he could produce a spark from a second winding on the same core if the current flowing through the first (or primary) winding was abruptly stopped. If he increased the number of turns in the second winding, the spark became stronger. In this way he created the concept of the transformer and the essential form of the auto ignition coil [6].

In 1832, in recognition of his achievements in electricity, he was offered the chair of "natural philosophy" (physics) at the College of New Jersey, now known as Princeton University. There he continued his research on electromagnets. One of his favorite demonstrations was an electromagnet that struck a bell when a battery was connected. He wanted to find out how long the wire could be made. At first he was able to strings it a few times around the lecture hall. Later he achieved a distance of 2.4km (1.5 miles). At Princeton he had a wire strung from the classroom to his house. Before a hushed audience he pressed the button, when his wife answered by pressing her button (after a few minutes), the students applauded [24].

In 1835 he invented the electromagnetic relay. This simple device was to assume enormous importance. He affixed an arm to his electromagnet with an insulated, U-shaped wire attached to it. When he connected the electromagnet to the battery, the arm dipped the ends of the wire into two mercury cups and thus closed a second and entirely different electric circuit [6]. In this way the signal could be refreshed, starting with a new battery and a new line and thus could reach any distance desired.

To his students at both Albany Academy and Princeton Henry was the ideal teacher. As one student put it: "He had a frank, open, commanding and dignified presence with an unaffected simplicity of manner inviting familiar and unconstrained intercourse"[24].

In 1846 Henry accepted the position of Secretary of the Smithsonian Institution and moved to Washington. (James Smithson was the illegitimate son of the Duke of Northumberland and a fellow of the Royal Society. Because of his illegitimacy he was rejected by English society; he moved to Paris and left his money to a nephew with the provision that, should the nephew die without an heir, the money should go "to the United States of America, to found at Washington, under the name of Smithsonian Institution, an establishment for the increase and diffusion of knowledge among men." The nephew died without heirs; for nine years the politicians in Washington bickered until finally an act was passed and a board of trustees was elected. Congress also commissioned a building (the "castle") and hoped that it would set the style for the city. Fortunately it didn't[6].

When Henry arrived in Washington the Smithsonian Institution was anything but glamorous. The location of the building was in an open, muddy field near a canal (which, in fact, was an open sewer and for years filled the building and its surroundings with its aroma). At the opening celebration wooden planks had to be laid so the visitors would not sink into the mud.

Henry was an able administrator of the Smithsonian Institution, though all his life he found it difficult to come to decisions. When Lincoln was elected President, Henry was bitterly opposed to him. Lincoln in turn did not think much of Henry or of the Smithsonian Institution; he was of the opinion that a lot of useless information was being printed there. But when the two men met they found that they liked each other and Henry became a close personal advisor to the President[6].

Henry died in May of 1878 at age 80. For his funeral the White House, Congress and several federal buildings were closed and the president, vice president and many in Congress attended. Later the U.S. House of Representatives and the Senate held an extraordinary joint session in memory of Joseph Henry, joined by President Rutherford B. Hayes[24].

And now we come to one of the strangest figures in our story: Henry Cavendish. He lived earlier than any of the people in this chapter, but his most important discoveries did not come to light until much later, simply because he never published them. Thus Cavendish presents a problem: should he be discussed with his contemporaries, even though his work had no influence? As a compromise, strange Henry Cavendish gets his own section here, midway between his life and the publication of his research.

Cavendish was born into nobility in 1731. His father was the son of the Duke of Devonshire, his mother the daughter of the Duke of Kent. At 18 Cavendish went to Cambridge, stayed there for four years, but didn't graduate. After Cambridge he moved back into his father's house in London. His father, Lord Charles Cavendish, was a Member of Parliament and a skilled experimenter in science. The younger Cavendish experimented, too, and soon surpassed his father [7].

Cavendish was an odd man, to say the least. He had a high-pitched, shrill voice and was morbidly shy. His father would take him to the social gatherings of the Royal Society; it was the only social activity Cavendish had. He would hardly ever participate in any discussions. If spoken to he would blush, stammer, squeak and disappear. Once he was introduced to an Austrian gentleman, who had come to London especially to see him. In the introduction Cavendish's achievements were praised in a pompous manner. Cavendish said nothing, turned around, darted through the crowd to his carriage and left [43].

He wore his clothes in a style two generations old, usually faded purple, with a three-cornered hat [4]. He shuffled around at the gatherings of the Royal Society in a quick, uneasy step, listening over the people's shoulders to what was being said, but disappeared when one of the participants noticed him. The senior members of the Royal Society passed the word that Cavendish was not to be spoken to or even noticed unless he spoke first. Even then, nobody was to look at him directly [7].

He was kept on a small allowance during his father's life, but when he was about 40 he inherited a large fortune. From this point on he lived in three houses, one of them he converted into a library, the other two into laboratories. The few visitors he had were startled to find such things as a forge in one of the rooms or a row of chamber pots used to hold crystalline solutions. He had servants but communicated with them mainly through notes. He was deathly afraid of women; if a female servant met him accidentally in the hallway, she was fired instantly [3].

Even though he never took any interest in financial matters and was annoyed by bankers who wanted to know what to do with his money, his fortune grew to more than one million pounds [43].

His world was science; nothing else interested him. But within science his expertise became far ranging, covering meteorology, chemistry, mechanics and electricity. He was the first to determine the composition of water. He accurately figured out the average density of the earth (5.448 times that of water) and, with its size already known, was thus able to calculate the weight of the earth [7].

He published only two papers on electricity, in 1771 and 1776. In the first paper he elaborated and provided the mathematics for Franklin's theory of electricity. The second paper is the more interesting one; it deals with the electrical properties of the torpedo fish. Cavendish had constructed an artificial torpedo from wood, covered with leather. Its electrical organs were made from pewter, with the electricity supplied by a Leyden jar. With this artificial fish he measured how much electricity was necessary to receive the same kind of shock as that delivered by the real fish. To do this he had to get an idea of how well water conducted electricity. In the paper he mentioned only that an iron wire conducts electricity 400 million times better than distilled water [5].

The part of the work done for this paper, which he did not publish, was in fact far more important. Hidden behind that brief statement is a complete discovery of Ohm's law, achieved without the use of any instrument to measure current. He had filled long glass tubes with various solutions, with wires stuck through corks on both ends. Then he sensed the shock from a Leyden jar, delivered through conducting fluid in the glass tube to his elbow or wrist. He pushed the wires further and further into the solution until the shock he received felt as strong as the one he received through a standard length of wire. In other words, Cavendish used his body as a galvanometer [7].

In this way he was not only able to determine the relative resistance of various media (such as distilled water, rain water or salt water) but also the relationship between the diameter of the tube, the distance of the wires and the resulting resistance. Fifty years before Ohm painfully unraveled his law (using much more advanced methods and instruments), Cavendish had arrived at the precise relationship. But he never bothered to tell anyone.

The discovery of Ohm's law before Ohm alone would have secured Cavendish a high place in science. But there were other discoveries of equal significance, again none of them published. He anticipated Coulomb by a few years and, in chemistry, he found Dalton's law of partial pressure before Dalton and Charles' law of gasses before Charles. Why he only published the unimportant part of his research is not known. It appears he had in mind to write a book some day, but never did. Perhaps he simply lacked the courage. Or perhaps he simply lost interest in the subject. When he died in 1810 he left 20 parcels of notes. James Clerk Maxwell, then a professor at Cambridge, edited the notes and published them in 1879 [5]. It was too late for them to be of any use, except for historical curiosity.

Much Ado About *Almost* Nothing

4. A Very Big Ship

A painter who promoted the telegraph
and a humongous ship that saved the transatlantic cable.

Since 1794 there had been in operation between Marseille and Paris a signaling system, which the French called the telegraph or the semaphore. It consisted of 120 towers placed on hills some 10km (6 miles) apart [7]. On top of each tower there was a huge wooden pole with a movable crossbeam, which in turn carried two movable arms at its ends. The crossbeam and the end arms could be placed at various combinations of angles through pulleys and strings from the tower below, giving 49 distinct signals. The operators read the signals through telescopes. A message of 50 letters took about one hour to transmit over the 600km (375 mile) distance, nearly 100 times as fast as a rider on a horse [8]. Although the system worked only during the daytime and was often shut down due to bad weather, it gave Napoleon a large advantage in his campaigns [9].

In England, a somewhat different system came into operation a year later between London and Portsmouth. It had six shutters, which could be opened and closed alone or in combination, giving a total of 64 different signals. A few years later such a system also connected New York and Philadelphia. In 1800 a semaphore line was installed between Martha's Vineyard and Boston so that the merchants could get an early report on the arriving ships and their cargo [7].

The idea of transmitting messages by electricity had surfaced even earlier than the mechanical telegraphs. In 1753 an anonymously published letter in a Scottish magazine proposed to string 26 wires, one for each letter. At the end of each wire there would be a pair of pith-balls, which would move apart when the line was charged with a Leyden jar and thus announce the transmitted letters in sequence. It was an attractive idea, for all anybody knew the transmission of electricity was instantaneous and such a system would not be shut down because of poor visibility.

The problem with this scheme was in the insulation. Stringing wires over a long distance brought with it a lot of leakage to ground, especially in a humid climate. In 1812 Francis Ronalds solved the problem by putting 175 meters of wire inside a series of glass tubes buried in the ground. He also had another innovation: two synchronized clocks at the ends, with the 26 letters of the alphabet marked on their faces. The sender applied a voltage to the wire when the clock showed the letter he wanted to transmit. The receiver wrote down the letter shown on his clock whenever his pith-balls parted. Ronalds tried to sell his telegraph to the British admiralty, but the conclusion of the military minds was that it was "not necessary".

With the advent of the battery and the electromagnet the possibilities for an electric telegraph became readily apparent. In 1831 Carl Friedrich Gauss and Wilhelm Weber set out to study the magnetism of the earth to find the underlying mathematical relationships. To do this, they had to make hundreds of delicate measurements of the strength and direction of the magnetic field and its variation with time, which made it necessary to make simultaneous measurements in two different places. To accomplish this, Gauss and Weber strung a wire over the rooftops of Göttingen, connecting the observatory with the laboratory of the university. They developed a receiver with an electromagnet and transmitted the data of the observation through a simple code, but the arm of the electromagnet was so heavy that it took over a minute to transmit a single letter. Gauss tried to interest the government in the invention but had no luck [7].

Professor Carl August Steinheil of the University of Munich came to see Gauss and Weber in the fall of 1835 and took over the development of the telegraph from them. With financial aid from the state he tried out several systems along railroad tracks and found that he could use the earth (or the rails) as a return path for the current, eliminating one of the two wires. He elaborated on the code, but he too couldn't find any buyers.

William Fothergill Cooke was not aware of any of these events. He had seen Steinheil's telegraph, but it must have been a rather primitive early version. Cooke thought that the idea of transmitting messages over an electrical wire was original with him.

He heard of the electromagnet and went to see Faraday, who gave him some encouragement but, when Cooke mentioned that he also had an idea for a perpetuum mobile, Faraday suddenly found he was too busy to help him. And Cooke needed help, because he knew very little about electricity.

Cooke was a promoter. He saw the possibilities of an electric tele-graph very clearly and he perceived that the railroads were the most likely initial users, for they were the only organizations that possessed long, continuous strips of land. Before he had his telegraph working he went to see the directors of the Liverpool and Manchester Railway. This railway, like other early railways, had a problem. The locomotives burned coke and spewed hot cinders. The ends of the lines naturally terminated in heavily populated areas and the residents of these cities were not too keen on having hot cinders drop on the roofs of their houses. For this reason the locomotive was shut down at the edge of town and hauled in by a long rope, wound by a stationary steam engine. In the case of Liverpool the situation was particularly critical: the rope went through a two-kilometer (1.4 mile) long tunnel. A signaling system was clearly necessary to tell the terminal when the rope was ready to be pulled.

Cooke was sure that the Manchester-Liverpool line was an ideal customer for his telegraph. But he came too late; the company had already decided on a pneumatic telegraph. This high-sounding name stood for a two-kilometer long pipe with a whistle at one end and a pump at the other.

Undeterred by Faraday's brush-off and by his first marketing failure, Cooke worked on his telegraph. But he ran into technical problems that he could not solve. So he went to see a professor at King's College in London by the name of Charles Wheatstone.

Wheatstone had inherited a musical instrument shop in London, but his passion was experimentation. Without any formal training in science he began to investigate the transmission and the properties of sound. He devised the "enchanted lyre", a striking demonstration of the transmission of sound through solids. Drilling a hole through the ceiling, he connected two lyres with a thin piano wire. A person on the upper floor would play the lyre, but the people assembled on the lower floor heard the sound from the lyre hanging on the nearly invisible wire.

In 1823 the 21-year-old Wheatstone wrote a paper on the transmission of sound, which was read before the French Academy of Science. In 1827 he built the "phonic kaleidoscope", a rod with a silver ball at the end which, when vibrated by sound, displayed the sound in figures. He was invited to publish this invention in the Journal of the Royal Institution and followed it with a second publication on the conduction of sound in the bone structure.

In 1833 Wheatstone performed a stunning experiment, which proved him to be a scientist of the first rank: he measured the speed of electricity. Wheatstone's approach is one of the truly great ideas in science because it is so simple. He strung a long wire so that its beginning, middle and end would be close to each other on a table. These three points in the wire all had spark gaps and he viewed their sparks in a rapidly rotating mirror. As the mirror continued to rotate between the first spark and the second and again between the second and the third, the images of the sparks showed up some distance apart. The faster the mirror rotated, the larger was this distance. Wheatstone even affixed a small siren to the mirror so that he could calibrate the rotational speed with the pitch of the siren.

A year after this last experiment, Wheatstone was appointed professor of experimental philosophy (which meant physics in those days) at the recently founded King's College in London. He stayed there to the end of his life. Rarely giving lectures, he devoted his time to research.

When Cooke came to see Wheatstone in 1837, Wheatstone had already experimented a little with telegraphic ideas. He looked at Cooke's half-completed apparatus and suggested an instrument with 5 wires and 4 magnetic needles. Four of the wires were connected to four coils driving the magnetic needles; the fifth one was the common return wire. Without an electric current the needles pointed straight up. When a battery was connected to a wire, the needle deflected to the right or left, depending on the polarity. The needles could either be actuated alone or in combination, giving 12 signals for a 5-wire, 4-needle system and 30 signals for a 6-wire, 5-needle telegraph.

Cooke suggested a partnership and the two men agreed. They were unlikely partners. Cooke was an aggressive promoter and speculator, Wheatstone was a conservative scientist.

They built a system based on Wheatstone's idea and found a railroad to pay for a short stretch of trial installation. It was a successful demonstration, but the railroad did not buy it. In 1838 they managed to convince the Great Western Railway to pay for an installation between Paddington and West Drayton, a distance of 21km (13 miles). Cooke took on the role of construction superintendent. Six wires were laid in iron tubes underground and the installation took a year to complete. It was an unqualified success. It not only helped the scheduling of the railroad, but started to transmit police and commercial messages as well. It was the world's first full-fledged electric telegraph system.

Cooke and Wheatstone moved on, mostly under Cooke's relentless drive. The second installation was for the London and Blackwall Railway, one of the strangest railroads of all time. The London and Blackwall was a single-track commuter railway, pulled by a rope, with a steam engine at each end of the track. There were seven stations, with one carriage at each station. The carriage at the end station clamped on to the cable first. When the conductor of the carriage in the second station saw the last carriage coming, he clamped his carriage to the cable just ahead of it. This process was repeated at each station until the all the carriages formed a train.

The problem was that, sometimes, a carriage in the middle of the system had not boarded all of its passengers when the end carriages appeared and inevitably crashed into it.   With Cooke and Wheatstone's telegraph each station could be forewarned of the oncoming carriages.

Soon there were additional installations, fueled by increasing publicity. The Great Western Railroad had a telegraph installed between Paddington and Slough.   On August 24. 1844 the Paddington station telegraphed to Slough that a known thief named Fiddler Dick had been seen boarding the train, third compartment, first second-class carriage.   As the train arrived in Slough, police officers opened the compartment door and asked if any of the passengers had anything missing.   A lady found her purse gone which, naturally, was found on Fiddler Dick, who was thunderstruck by the supernatural discovery.

An even more startling demonstration of the powers of the telegraph occurred on January 2, 1845.   A woman was murdered at Salt Hill and a man was seen hurrying from her house in the direction of Slough station.   The police telegraphed a description to Paddington and there the suspect was shadowed by the law and arrested the next day.   The news of how the arrest came about was like money in the coffers of Cooke and Wheatstone.

The business of the partners expanded rapidly and profitably, but they started to quarrel.   Cooke suddenly claimed that he, not Wheatstone, was the main inventor.   Wheatstone, who was much more interested in science and development than management, designed an advanced telegraph system, consisting of a small rotating head which showed letters directly and used only a single wire.   It used a principle similar to the later rotary dial of the telephone, advancing the head through a series of pulses.   A year later he made this little machine not only show the letters but print them as well.

The tension between the partners grew worse as their company became more and more successful.   They began to hate each other.   In 1845 Cooke found backers, bought Wheatstone out for £ 30,000.   By 1870 he was in deep financial difficulties.   He died, broke, in 1879.

Wheatstone went on doing original scientific work in acoustics, electricity and optics for the rest of his life and came up with several additional inventions, such as a typewriter and perforated paper tape for storing data.

Wheatstone was no recluse scientist.   He was a shrewd businessman who wisely invested the money he received from the telegraph and many of his other inventions.   When he died in 1875 he left a substantial fortune.

In 1832 the sailing ship "Sully" returned from Europe to New York. She was a small and slow ship; the trip took six weeks.   Among the passengers was a Dr. Charles Jackson, a physician from Boston, and the painter Samuel Finley Breese Morse. The two were having a good time together, talking about what they had seen and experienced in Europe.

Morse, who was called Finley by his family and friends, was born in 1791 and grew up in Charlestown, Massachusetts, the son of a minister. He went to Yale and, at 20, traveled to Europe to study the masters. After his return he opened a studio in Boston, but didn't find enough work. He became an itinerant painter, got married, had three children and settled in New York [10].

The struggling Morse was beset by misfortune; his wife and parents died within a short time of each other. Forlorn, he placed his children with relatives and went to Europe a second time, where he stayed for three years.

When he returned on the Sully, Morse was 41 years old. He was a good painter but had never been commercially successful. He had managed to squeak through financially, but his prospects for the future were dismal and he saw no possibility to take his children all under one roof.

The dinner conversations with Dr. Jackson and the other passengers were pleasant. They talked about the latest European fads and inventions and inevitably electricity became the main subject. Jackson was the type of man who knew something about everything, though without understanding much. Morse had some knowledge about electricity, but not nearly enough to be competent in it. Jackson described the newfangled electromagnet. The party wondered how far electricity could move along a wire. Jackson confidently predicted that, since the propagation of electricity was instantaneous, there was no limit [11].

These remarks reached a flashpoint in Morse's mind. He began to sketch a telegraph system in his painter's notebook. Short and long electric signals - dots and dashes - actuated an electromagnet. On its arm the electromagnet carried a pencil and the pencil drew the signals on a moving strip of paper in the form of a wavy line [11].

His first idea for a code was cumbersome. It consisted only of numbers: one dot for a one, two for a two, etc., up to numeral five. Six, seven, eight, nine and zero were the same, except that a dash was added after the dots. With this code he could transmit any decimal number and he proposed a large dictionary in which every word was assigned a number [11].

Stepping off the Sully in New York he was full of enthusiasm and talked about nothing but his telegraph. Like Cooke, he believed himself to be the first one ever to think about transmitting intelligence over an electric wire. For the next few days he experimented incessantly in his brother's house, but nothing worked. He simply didn't know enough about electricity.

Morse badly needed an income, so he found a job as professor of arts and design at the University of the City of New York. It wasn't much of a job; he got paid according to the number of students he was teaching and they were few. Off and on he continued tinkering with the telegraph, but his efforts were pathetic. He wound the copper wire directly onto the iron core without insulation, thus shorting out the coil.

In the meantime Morse became involved in politics. Beginning about 1820 large masses of foreigners, mostly Irish Catholics, arrived in New York. Irritated by the unkempt, unmannerly immigrants who worked for low wages, stealing the jobs of native-born Americans, he, a Protestant, wrote enraged pamphlets against an imagined secret plot of the Roman Catholic Church to take over the United States. In 1836 he ran for mayor of New York, but got less than 5% of the votes [10,12].

Despite this extracurricular activity he continued working on his telegraph. In January of 1836 he showed his apparatus to a colleague at the university, professor Leonard Gale of the chemistry department. Gale had a better knowledge of electricity and he brought to the attention of Morse the work done by Joseph Henry on electromagnets. Morse went to see Henry, who gave him encouragement and a few helpful hints. With the assistance of Gale, the telegraph finally began to work. Morse wrote out a caveat, the first step then used in obtaining a patent, and sent it to Washington.

Samuel F.B. Morse

In September of 1837 Morse demonstrated his telegraph at the university over 200 meters (600 feet) of wire. He was still using only numbers but he was clearly able to transmit messages. His colleagues were impressed and so was a young man, a former student of his named Alfred Vail. Vail's father was the owner of ironworks in Morristown, New Jersey. They struck a deal with Morse and Gale, whereby Vail became a partner and assistant and his father put up $ 2,000 [11].

Luckily for Morse, Alfred Vail was a skilled mechanic and had a sharp, inventive mind. Together they improved the mechanism and changed the code. Instead of using numbers and a dictionary, they now assigned combinations of dots and dashes to numbers and letters. In this way the dictionary was no longer necessary; all the operator had to know was 36 different combinations of long and short signals. Under the agreement Vail had signed, all inventions belonged to Morse and were to be patented in Morse's name only.

To choose the combinations they used a printer's type-box. The letters which contained the most types (and were thus presumably used most often) received the simplest codes [12]. It was a brilliant idea, although the "Morse" code has been changed twice since then by international agreement and only four of the signals are still the same.

Years before, the U.S. government had offered $ 30,000 to anyone who could build a telegraph system workable over a thousand miles up and down the Atlantic coast. What the government had in mind then was a semaphore system, but the offer was never taken up. Morse came across the half-forgotten resolution and decided to appeal to Congress for financial aid.

He took his apparatus to Washington and demonstrated it to the House Committee on Commerce and to President Buchanan. But the country was in a serious depression and money was hard to get, especially for such a far-out thing as an electric telegraph [12].

The chairman of the Committee on Commerce, congressman F.O.J. Smith from Portland, Maine, knew a good thing when he saw it. "Fog" Smith, as his colleagues called him with a slight change of his initials, was 15 years younger than Morse, yet had already acquired quite a reputation for double dealing and shady business ventures. He talked Morse into becoming a partner and Morse agreed to form a partnership with Smith, Vail and Gale [11].

Without resigning from Congress, Smith had a bill introduced to appropriate $ 30,000 for the construction of a test line. Then he took a leave of absence and sailed with Morse to Europe to take out patents there. In England Morse was coldly rebuffed. Cooke and Wheatstone's telegraph was already in operation; without examining Morse's apparatus the Attorney General rejected the application [12].

In France the reception was friendlier; Morse's telegraph was demonstrated before the French Academy of Sciences and he was able to apply for a patent. But when he tried to sell the rights to the French government there was no reaction. Under the French law he now had to bring his invention to commercial feasibility within two years, but when investors were rounded up to build a test line, the government blocked the effort by claiming the operation of telegraphs was a state monopoly [12].

Morse spent a year in a fruitless effort to obtain patent protection in Europe. In the U.S. the appropriations bill was stuck in committee.

Discouraged and broke, Morse returned to New York. His patent was issued in 1840 but the situation looked hopeless. All Morse wanted for his telegraph was to sell it to the government so that he could continue painting without having to worry about money. He had no faith in private enterprise and strongly felt that an important means of communications such as the telegraph should be the exclusive property of the government. But Congress didn't see any great importance here. For four sessions no action was taken on the bill. "Fog" Smith had run for governor in Maine and lost; he was now no longer in Congress. In addition he got himself into financial difficulties and was of no use to Morse anymore. Although desperate for money, Morse spent months in Washington trying to persuade lawmakers. Finally, in early 1843 the bill came to the floor. Most of the members of the House at first thought it was a joke. Finally, on March 3, the last day of the session, $ 30,000 was appropriated for the building of a telegraph line between Baltimore and Washington [11].

Morse was now the Superintendent of United States Telegraphs, at a salary of $ 2,000 per year. He hired James Fisher, a colleague from his university, as well as Vail and Gale [11].

The line was to be laid underground in a tube of lead, using the right-of-way of the Baltimore and Ohio Railroad. "Fog" Smith, through a straw man, awarded himself the construction contract.

Morse and his colleagues let it be known that they were looking for a special plow, which could be used to bury the cable in the ground. Ezra Cornell, who was then working as a plow salesman in Maine, heard about it and came up with a design, which in one step opened the ground, buried the cable and closed the ground behind it. He was hired as an assistant. Cornell, a lean and hardworking son of a farmer and potter from upper New York state, was a levelheaded, multi-talented fellow. It occurred to Cornell that the cable should perhaps be tested periodically. On his own initiative he made a test after a few miles were laid and the cable turned out to be a complete short-circuit. The hot lead drawn over the insulation had burned it and the cable had to be pulled out of the ground again. Fisher, who had been responsible for the manufacture of the cable, was fired. Smith, in the meantime, had managed to spend $ 20,000. There was no way the line could now be completed with the underground approach within the budget [11].

Cornell suggested that the wire be carried by poles and drew up a plan. In his approach glass doorknobs were used to insulate the wires from the wooden poles. Vail objected violently to Cornell's idea. Morse, unable to make a technical judgment, went to see Joseph Henry, who backed Cornell. From this point onward progress was rapid and by May 1, 1844 the line reached Annapolis Junction. Luck in the form of an unexpected public relations "spectacular" came Morse's way. The Whig-party convention was being held in Baltimore. The operator at Annapolis Junction, gathering the news of the convention from passenger trains, telegraphed it to Washington [11].

On May 24, **1844** the 65 km (40 mile) line was complete and Morse staged a demonstration for Congress. It was at this time that Morse transmitted from a chamber of the Supreme Court (then inside the capitol building) the famous message "What hath God wrought?" [11].

Morse's public-relations "bonanza" continued. On May 25, the Democratic convention opened, also in Baltimore. The telegraph terminal point was now in a room below the Senate chamber and Morse kept delivering news bulletins from the convention, which were read aloud on the Senate floor. The convention nominated Polk as president, which came as a complete surprise to the senators. It then nominated Silas Wright for vice president. Wright, however, was not at the convention but in Washington and sent his refusal to Baltimore by telegraph. The incredulous conventioneers sent a delegation by train to confirm the news [11].

Within a few days Morse opened the line for other uses. The police succeeded in apprehending a criminal, trains began to be dispatched via telegraph and commercial messages were sent. But for the ordinary person this marvel, which sent messages at lightning speed was a bit difficult to comprehend; Morse received frequent request to send packages by telegraph.

Morse was now riding high; he had a second bill introduced in Congress for the government to buy the rights to the telegraph. He was asking $ 100,000 for himself. But Congress did nothing and Morse reluctantly formed the Magnetic Telegraph Company to build and operate telegraph lines. Fortunately, he hired a capable business agent, without whom he would almost certainly have failed.

The Magnetic Telegraph Company prospered. Morse's system had an unexpected advantage over Wheatstone's telegraph. As conceived by Morse the dots and dashes were to be recorded on a paper strip. But it turned out that his operators, after only a few weeks of practice, were capable of "hearing" the letters and numbers directly from the clicking of the electromagnetic receiver. In this way the transmission speed was limited only by how fast the operator was able to write.

By 1847 Morse was able to buy a 100-acre estate on the Hudson River south of Poughkeepsie. At 57 he married again and had his children, now grown up, in his house for the first time in 20 years. But dark clouds kept appearing for Morse. Dr. Jackson from Boston claimed that he, not Morse, had invented the telegraph on the Sully. Jackson was an easy case compared to the dozens of rival systems that suddenly began to appear once the telegraph was a reality. Battling his rivals kept Morse busy in court [11].

Morse was a bitter fighter. He felt that the telegraph was his and his alone. Unfortunately Vail had published a book on the history of the telegraph shortly after the completion of the Washington - Baltimore line and in this book he ignored Henry's contribution. Henry felt hurt and let it be known that he had helped Morse and that much of Morse's success was due to his electromagnet [16].

The early patent system in the Unites States was considerably different from the one used now. The first step in obtaining a patent was to write a caveat, a preliminary description of what the inventor was working on, submitted merely to establish priority. After the invention was made, a patent application was written and submitted. If there were no prior claims, the Commissioner of Patent then issued a letters patent, with a life of seven years. After seven years the inventor could ask for his patent to be re-issued and include features, which had been added in the meantime. Morse did that. When his patent re-issued in 1846 it contained wide-ranging claims - including the electromagnet itself - so that any kind of electromagnetic telegraph was covered [11].

Did Morse purposely include inventions he knew he had not made? Henry pointed out that, as far as he knew, Morse had never made a discovery in electricity, magnetism or electromagnetism, that he simply combined the discoveries of other people in a working system. Morse, enraged by Henry's statement, wrote a pamphlet in his defense in which it becomes clear that he had begun to believe that he invented a great many things he certainly did not invent [15]. Fame had been working on his mind.

Despite Morse's patent claims and some decisive battles in court, the number of telegraph companies multiplied rapidly and competition became fierce and vicious. By 1849 there were three different systems between New York and Boston. The many competitors soon realized that they were hurting themselves and started to combine. Morse's company merged into the American Telegraph Company. By the beginning of the Civil War there were three companies left: the American Telegraph Company, the United States Telegraph Company, and Western Union. The first two had most of their lines running north to south, while Western Union's line ran east to west, north of the fighting and extending all the way to San Francisco. During the Civil War the north-south lines suffered substantial damage while the east-west lines remained intact. After the close of the war Western Union emerged the strongest of the three and was able to absorb the others [4].

Morse did reasonably well in all these transactions and he ended up making far more money than the $ 100,000 he had asked of Congress. But he was not happy. He fought bitterly with all of his associates and broke with

Gale, Vail and Smith.

With money and more leisure on his hands he started renewing his old racism again, this time on the side of slavery. He lashed out at the abolitionists by quoting justification for slavery from the Bible. He ran for Congress in 1854, but lost. His pamphlet writing continued until his death in 1872 [12].

Of the original people around Morse, Ezra Cornell was perhaps the most remarkable. Hardworking as ever, he continued building telegraph lines and investing the profits in telegraph stock. Spending no time on lawsuits, he became gradually much richer than Morse and financed the construction of a large public library and model farm in Ithaca, New York. Subsequently he moved into politics, becoming an assemblyman and senator in the New York state legislature. His library and model farm, together with a gift of $ 500,000, became Cornell University in 1868.

Things had gone wrong for Frederick N. Gisborne. He had started out with a great idea: a telegraph cable connecting the tip of Newfoundland to the growing telegraph system of Canada and America, so that messages could be dropped off ships in transit from Europe and arrive in New York two weeks earlier. Gisborne obtained a concession from the Canadian government, raised money and began to build the line. But he had underestimated the cost of crossing the 600 km (375 miles) of rough wilderness in Newfoundland and he ran out of money [4].

In 1854 he was in New York looking for help. Among the people he talked to was Cyrus Field a 34-year-old, self-made businessman. Field listened to Gisborne's troubles, but then came up with an idea of his own. A Newfoundland to New York connection had only a slight advantage; a really big step would be a connection - underwater - from Newfoundland to Ireland, so America could communicate with Europe. Field quickly fired off two letters, inquiring if such an undertaking were technically feasible. The first one went to Morse, who answered: "yes, no problem." The second letter went to the superintendent of the Depot of Charts and Instruments in Washington (later the U. S. Naval Observatory), who had started charting the ocean currents and compiling topographical maps of the ocean floor. The opinion of the superintendent was that such a venture appeared feasible. Field found out that, between Newfoundland and Ireland, the Atlantic Ocean showed a plateau with a maximum depth of 4 km (2.5 miles) with a covering of small shells. Not only was this the shortest distance, 3'300 km (2,000 miles), but it also was the most suitable place to lay a cable [4].

These two answers were enough for Field to move. He bought out

Gisborne and formed the New York, Newfoundland and London Telegraph Company, raising a total of $ 1.5 million. He put his brother Matthew in charge of building the line through Newfoundland. Matthew left and hired 600 men; the terrain was so rough the entire crew had to be supplied from the sea [4].

The line from Newfoundland to mainland Canada had to cross 137 km (85 miles) of the Gulf of St. Lawrence. Cyrus Field went to England to order suitable cable. There he met John W. Brett who had laid the first submarine cable across the English Channel. Brett had met with mixed success. He had used a newly discovered insulator, gutta percha, around a copper wire. Gutta percha, similar to rubber, is obtained from the sap of a tree, but unlike rubber, it is fairly hard at room temperature and does not deteriorate in saltwater. Only one day after Brett had his cable across the channel a fisherman pulled it up with his anchor and, thinking that it was a strange seaweed with a core of gold, cut out a section. A year later Brett laid a second cable, again insulated with gutta percha, but this time he added a layer of hemp and a protective shield of galvanized iron wires. This cable worked, but the transmission of signals through it was annoyingly slow [4].

Brett's efforts had laid the foundation for a submarine cable industry in England and Field found a ready supplier for the first section. The cable arrived in the Gulf of St. Lawrence the following spring and began to be laid by a sailing ship pulled by steamer tugs. But it soon became clear that the maneuverability of a sailing ship was too poor for the job and the laying was abandoned. To lay a cable across the Atlantic, steamships would be essential [4].

Cyrus Field

Field hurried back to England to order replacement cable. The second attempt to cross the Gulf of St. Lawrence was successful - using a steamship - but the line from St. Johns, Newfoundland to New York had taken two years and consumed $ 1 million of the capital [4].

Field went again to England, this time to raise more capital. He convinced the British government to guarantee a £ 350,000 investment by the public and provide two naval steamships for the cable laying. He formed the Atlantic Telegraph Company and merged the interests of the previous investors into it. He then went back to America and obtained an identical guarantee and the loan of two Navy steamships from the U.S. [5].

Now Field was ready to order the cable and organize the laying. He was impatient, the preliminary work had taken too long, and he allowed only six months for the manufacture of 4,000 km (2'500 miles) of cable, not enough time to test it or to practice laying it.

But there were other nagging questions too. Both of Brett's cables across the channel allowed only slow transmission and the signal arrived considerably weakened. How long would transmission across the Atlantic take? Would the signal arrive at the other end at all?

Field had hired Charles Bright as chief engineer and a Dr. Whitehouse as electrician. Whitehouse recommended a cable 1.7 cm (0.7 inches) in diameter. It consisted of a copper wire, three layers of gutta percha, a layer of hemp saturated with tar, linseed oil and wax and outside armor of iron wire. Faraday was consulted on the question of transmission delay in the cable and, after experimenting with short lengths, he came to the conclusion that it would take one to two seconds for the signal to travel the entire length [5].

But William Thomson, professor at Glasgow University, had some doubts. At the time Field met him and invited him to become a director of the Atlantic Telegraph Company, Thomson (later Lord Kelvin) had acquired an international reputation as a mathematician [18]. Until his encounter with Field, Thomson's work had been entirely theoretical, but he had studied the behavior of a cable; he likened it to charging a huge Leyden jar through a long, thin wire and he recommended that the size of the wire and the diameter of the cable be increased and that copper of the highest purity be used [18].

But Whitehouse was not impressed by the abstruse theories of the professor and, besides, it was too late; the cable was already being manufactured [4].

On May 14. 1857 the 5,200 ton U.S.S. Niagara, the largest steam frigate in the world, arrived in the Thames estuary and there she was joined by the 3,200 ton British warship Agamemnon. Each ship was to carry 2'100 kilometers of cable; it took three weeks to stow the two halves. The two screw-propeller ships were to be accompanied by the Susquehanna, the largest side-wheel steamer in the U.S. Navy, and the Leopard, a paddle steamer from the British Navy [5].

William Thomson
(Lord Kelvin)

The ships met in Valentia Bay on the West Coast of Ireland. Dr. Whitehouse was supposed to be on board to supervise the performance of the cable, but he pleaded ill health at the last minute and William Thomson was asked to replace him. The group of ships started westward. On August 10, after laying 560 km (350 miles), the cable parted. The ships returned to Valentia Bay. Since new cable needed to be manufactured to replace the lost portion, it was decided to postpone the expedition to the following summer and in the meantime improve the paying-out mechanism. The cable was unloaded and stored on shore [1].

The speed of transmission through the full length of cable had been found to be a disappointing two words per minute and the receiving apparatus provided by Whitehouse was so clumsy that, even with 500 batteries at the sending end, it would not work reliably. At two words per minute it was questionable whether the cable would ever pay back its investment. The directors decided to give Whitehouse the assignment to try to improve transmission speed by having the operators practice during the winter, using the actual cable stored on shore.

But William Thomson was still skeptical. His calculations told him that Whitehouse was going about it the wrong way. In order to detect the weakened signal faster, a more sensitive receiver was needed. He, the theoretical mathematician, proceeded to construct one and he asked the board of directors to allocate £ 2,000 for the project. The directors thought this was far too much and cut it down to £ 500. Using mostly his own money, Thomson built a pair of instruments. Instead of using a heavy electromagnet, the current from the cable moved a delicately suspended coil, to which Thomson affixed a mirror. A paraffin lamp shone its light through a slit onto the mirror and as coil and mirror were rotated slightly by the arriving current, the movement of the reflected light could be seen, greatly magnified, on a screen. With this sensitive instrument Thomson was able to reliably transmit eight words per minute [18].

When the cable-laying party came alive again in May of 1858, Whitehouse again pleaded illness and Thomson stepped in, carrying with him his own instruments. The cable was stowed in the Agamemnon and the Niagara. Two new British support ships, the Gorgon and the Valerous were assigned to accompany the two larger ships. The group started out from Plymouth on June 10 for a rendezvous in mid-ocean. It had been decided to splice the two halves there and then lay them simultaneously in the opposite directions [5].

The four ships immediately ran into a heavy storm and for a while the Agamemnon was in serious trouble. Unable to stow all of the cable in her holds, a portion of the heavy load had to be carried on deck, which caused the ship to roll dangerously. The storm lasted a week and scattered the ships, but fortunately they all found each other and were able to splice the cable on June 26.

But the bad luck persisted. After only 10 km (6 miles), the cable parted on the westward-going Niagara. They met again, spliced the cable a second time and went off in their separate directions on June 27. After 129 km (80 miles) the copper wire in the cable ruptured somewhere on the bottom of the ocean [5]. Thomson had connected his mirror galvanometers and the Niagara and Agamemnon were in constant communication through the cable. Thus, when a break occurred on one ship, the other one knew instantly and returned to the starting point.

For the third time the cable was spliced and on June 29 the laying started anew. This time the ships managed to pay out 410 km (256 miles), when the transmission suddenly stopped again. In the meantime a dense fog had risen and when the ships returned to the mid-ocean point they couldn't find each other. Reluctantly they headed back to shore [5].

There was sufficient cable left to try again and it was still early enough in the season. On July 29 the four ships met again in mid-ocean and the cable was spliced. This time there were no breaks in the cable, except for an interruption of the transmission that mysteriously corrected itself. The Niagara arrived at Trinity Bay, Newfoundland, on August 5. 1858 and the Agamemnon the same day at Valentia Bay. The triumphant Field, who had been at sea with his ships on all six attempts, sent a telegram from Trinity Bay to New York, announcing the completion of the transatlantic cable. New York went wild. The stock of the Atlantic Telegraph Company, which had previously sold at $ 300, shot up to $ 1,000. In Boston a 100-gun salute was fired on the Common and the bells of the city rang for an hour. Harpers Weekly carried Field's portrait on the front cover. Charles Bright, the chief engineer, was knighted upon his return to Great Britain. The cable then carried an official message from Queen Victoria to President Buchanan and the President sent one to the Queen [5].

The official day of celebration in New York was set for September 1, with Field as the guest of honor. It began with solemn services at Trinity Church. A six-hour procession led from the Battery to the Crystal Palace at 42nd street. Speeches were given and medals pinned. At night there was a torchlight parade and a dinner for 600 guests and City Hall was lit by fireworks [2].

That very night the cable failed.

The operators and electricians had suspected for days that something was wrong. The transmission was intermittent and often garbled. The 98-word message from the Queen had taken more than sixteen hours to transmit

For days operators at both ends frantically tried to re-establish communications, but the cable remained silent. It was determined later that the cable's demise had started the year before, when it was stored on shore and exposed to the sun. The insulation had become brittle and when Whitehouse, in a vain attempt to make his apparatus work, had stepped up the voltage to 2,000 Volts, the insulation had begun to deteriorate. There was no possibility of repair; the cable was a total loss [4].

Any ordinary businessman would have given up at this point, but not Cyrus Field. He convinced the British government to increase its guarantee for a new stock issue. A technical committee was formed to decide on the specifications of a new cable. The new cable selected had a much thicker copper core (as Thomson had suggested) and had a diameter of 2.8 cm (1.1 inches). It was considerably heavier than the first cable but much stronger and, due to the increased insulation, was lighter in the water thus producing less pull on the payout mechanism [4].

But time was against Field. Civil war preoccupied the United States and no money could be raised nor ships obtained from the Navy. Undeterred, Field labored on to keep the great enterprise going. And suddenly luck came his way in the form of a giant ship, the Great Eastern. She was designed to hold enough coal to steam from England to Australia and back, around the tip of Africa. 230 meters (700 feet) long (with a double iron hull), she had a displacement of 32,000 tons, five times that of any other ship at that time. Two steam engines drove a gigantic stern propeller and two paddle wheels [6,14].

The year the Great Eastern was launched another great venture had begun: the Suez Canal. It would open in 1869 and cut a large portion off a voyage to the Far East and Australia for any ship, except the Great Eastern; she was too large to pass through the canal. After only a few passenger trips she was sold at auction for $ 25,000 [14].

The Great Eastern was precisely the right ship for laying a transatlantic cable. She could hold the entire length of the new, heavier cable with room to spare. Moreover, she was a steady, huge platform and the combination of paddle wheels and screw made her highly maneuverable. Field was able to lease her for very little money.

With the right ship available and the Civil War coming to a close, Field was ready to move. The Great Eastern was outfitted with three large holding tanks and the cable was carefully coiled into them (it took five months). A crew of 500 was assembled and supplies stored for several weeks of travel [5].

On July 15. 1865 the Great Eastern, again with Thomson and Field on board, moved to Valentia where the cable was connected to the shore.   On July 23. she headed west with an escort of two British warships.   After only 135 km (85 miles) of cable had been laid, transmission stopped and the measurements indicated that the cable was shorted out 16 km (10 miles) astern.   The cable was reeled in, the defective section was cut out, the cable spliced and the journey continued [5].

The Great Eastern

The same thing happened four days later, after another 1,000 km (625 miles) had been laid.   On August 2. another fault was detected.   This time the reeling-in did not proceed smoothly.   The ship drifted over the cable and chafed it with her iron sides.   As the cable was pulled up it parted and sank into four kilometers of ocean.   Field broke down and cried like a child.

For ten days the Great Eastern grappled for the cable.   Three times the cable was hooked but each time the rope broke under the strain of hauling up the cable.   The location of the cable was marked with a buoy and the Great Eastern returned to Great Britain [5].

There the stouthearted directors of the Atlantic Telegraph Company decided they would raise another £ 600,000 not only to grapple and splice the cable already on the bottom of the ocean but lay a second cable as well.   A new company was formed, the Anglo-American Telegraph Company, taking over the interests of the old one.

A year later, on Friday, July 13.   **1866** the Great Eastern left Valentia again, laying the second cable.   The trip was uneventful, the cable paid out smoothly and without interruptions.   Connection was made at Heart's Content, Trinity Bay on July 27.   The Great Eastern returned to the end of the first cable, grappled for three weeks with stronger gear and lifted it on September 1.   She arrived back at Trinity Bay with the second cable completed on September 7 [5].

It had taken Cyrus Field 12 years to complete the venture, which Morse had said would be "no problem", and he had to cross the Atlantic 64 times. The total investment was £ 2.5 million, or $ 12 million.   Transmission started at eight words per minute, gradually increasing with experience to 16 words per minute.

Alas, the two cables were only used for eight years.   In 1873 and 1874 two new and faster cables were laid.   After this time the first cables were relegated to backup status and by 1877 they were officially out of service. By 1880 there were nine separate cables connecting Europe with the United States, all laid with relative ease [9].

Cyrus Field made a large fortune from his stock in the venture.   He settled in New York and used much of his money to buy control of the New York Elevated Railroad Company.   By 1881 he was worth $ 8 million, had a 700-acre estate on the Hudson and a residence in town.   The New York Elevated merged with the Metropolitan Railway Company, controlled by Jay Gould.   Out of character, Field started to speculate in commodities on an 80% margin and helped Gould drive the Railway stock up.   But Gould was easily able to outmaneuver Field and Gould was no friend.   Simultaneously with a collapse in commodity prices in 1887 he staged a selling panic in the Railway stock and, within a 24-hour period, Field lost his entire fortune and only had left what was in his wife's name.   After this Field, who suffered from high blood pressure, became moody and withdrawn.   In 1891 his wife died and a few days later his son was arrested for embezzlement.   Field spent the rest of his money covering some of his son's debts.   Shortly thereafter he became more and more delirious, no longer recognizing his family; he imagined himself standing on the bridge of the Great Eastern, shouting commands.   He died in 1892 [2].

What happened to the Great Eastern?   After a few more years of laying cables between Europe and America and from Bombay to Aden she was refitted for passenger service, but was a commercial failure again.   The proud giant found herself anchored in the Mersey River and used as a showboat.

She was finally broken up and sold for scrap in 1898 [6]. Not until 1906 was there a ship that exceeded her in tonnage.

5. Mr. Watson, Come Here!

There were many inventors of the telephone,
but only one was successful.

In **1854** a Frenchman, Charles Bourseul, wrote a titillating paper: he proposed an apparatus to transmit speech via telegraph. His idea was to affix a contact to a membrane; as the membrane was moved back and forth by the sound waves, the contact would open and close. With a battery in the circuit the wire would thus carry pulses in the rhythm of the speech. At the other end the wire would be connected to an electromagnet mounted close to a thin iron membrane. This receiving membrane, being pulled toward the electromagnet in the rhythm of the speech, would then recreate the original sound pattern.

Charles Bourseul only wrote about his apparatus, he never built it. Johann Reis was different. The same idea occurred to him a few years later and he did something about it. Reis was a schoolteacher and had visions of becoming a scientist and inventor.

In **1860** he built an artificial ear out of wood with a membrane behind it and a delicate electrical contact attached to the membrane. As a receiver he used a violin. A steel knitting needle with a coil of electrical wire around it functioned as an electromagnet, vibrating one of the steel strings of the violin.

His idea worked - after a fashion. There was some sort of sound coming out of the violin, but it was badly garbled, making it impossible to recognize any words. He demonstrated his invention to various scientific circles, which found his "Telephon" curious and interesting. But it was far from being a practical instrument, since it was quite impossible to understand the sound coming out of the receiver.

Bourseul's and Reis' ideas had no chance of ever working. By converting the sound waves into crude pulses most of the information in the speech is lost. The only part of the original sound that remains is the fundamental tone or "pitch" and to this is added a large amount of distortion.

Alexander Bell (he adopted the middle name Graham at age 11 and was called Aleck by his family) was born in 1847 in Edinburgh, Scotland. His father was a teacher of speech articulation, as his grandfather had been. His mother was deaf and his father acquired some fame as the inventor of a system of graphic symbols, called "visible speech", which helped deaf people form sounds correctly [2].

Bell followed in his father's and grandfather's footsteps, becoming a teacher of elocution. Trained by his father, he had a finely honed sense of hearing (and perfect pitch).

In 1867 Bell's younger brother died of tuberculosis. Three years later his older brother succumbed to the same illness and Bell himself was getting ill. To save the only remaining son's life, the family decided to emigrate to a different climate, to Canada. They settled in Brantford, Ontario, near Montreal.

Before leaving Britain, Bell had come across a French translation of a book by Hermann Helmholtz, an eminent German physiologist who understood the ear (and could have told Reis that his "Telephon" would never work). Among other things Helmholtz described a tuning fork (or resonator) powered by an electromagnet, which emitted a continuous, pure tone. Bell read the book on the boat and was now getting enamored by an idea he had carried around in his head for years. He had noticed that the strings of a piano could be made to vibrate by singing into the piano. Only the string resonating with the particular pitch would respond, the others would be still [4].

What Bell had in mind was an improved telegraph. A wire could only carry one message at a time and some cities were already becoming densely forested with telegraph poles; there were so many overhead wires that the telegraph system threatened to become a hopeless, ugly mess. If there were some way to transmit two, four or even eight messages simultaneously, the savings of copper wire and installation costs would make the inventor a fortune. Bell's idea was to use several of Helmholtz's tuning forks to create different individual frequencies. These resonators would then be turned on and off with a telegraph key using the Morse code and their electrical signals transmitted together over a single wire. At the far end there would be an equal number of tuning forks, made to vibrate by an electromagnet. Only the fork tuned to the correct frequency would respond and thus several operators could receive messages by listening each to a different tuning fork [1].

The following spring Bell was invited to lecture at several schools for the deaf in the Boston area. In the fall of 1872 he moved there permanently and accepted a position as professor at the newly formed Boston University. He also had a private practice teaching deaf children to speak [4].

In Boston, Bell met two prominent citizens who would soon become important in his career. Thomas Sanders was a leather merchant from Haverhill, whose congenitally deaf son George Bell taught to speak. Gardiner Greene Hubbard was a lawyer from Cambridge, across the Charles River, who was active in public affairs. He was the first president of the National Geographic Society, a regent of the Smithsonian Institution and a member of the State Board of Education. His daughter Mabel had become deaf after she contracted scarlet fever at age five. Bell supervised her speech education [4].

Life was not easy for the professor; his pay was very low. After a year he moved into the house of the grandmother of little George Sanders in Salem and commuted by train. It was there, in his room in Salem, that he started to work on his invention. At first he used tuning forks but later he switched to vibrating steel reeds, made from clock springs [4].

But Bell couldn't get his apparatus to work. His manual dexterity was poor and his devices clumsy. He needed the help of a craftsman, but that took money. When he explained his idea to Sanders and Hubbard he found two ready sources for financial assistance. The three of them formed a partnership on a handshake. Each of the partners was to receive a one-third interest in the invention [4].

Now Bell could afford to have his transmitters and receivers made by a qualified mechanic. In the fall of 1874 he hired the services of the shop of Charles Williams at Court Street in Boston. The machinist assigned to the job was Thomas A. Watson. Watson was seven years younger than Bell, being 20 when they met for the first time. He was a quick-witted, personable young man who was one of Williams' best workers. He, too, lived in Salem and commuted to work daily by train [4].

Williams' shop made a variety of electrical devices: call bells, gongs, hotel annunciators, batteries, and telegraph transmitters and receivers. But the specialty of the shop was custom work for electrical inventors.

In October of 1874 Bell heard for the first time of Elisha Gray. Gray was 11 years older than Bell and worked at Western Electric Manufacturing Company in Chicago, the largest and by far the best supplier of telegraph equipment. That year Gray made an accidental discovery which appeared so revolutionary to him that he took a leave of absence from Western Electric to pursue it: he found that he could produce a sound by mounting an iron washbasin close to an electromagnet, with the latter powered through a rapidly opening and closing electric contact. This was, in fact, the first,

Hans Camenzind

primitive loudspeaker, though it had no other sound to reproduce than the tiring buzz of a vibrating contact. Gray had an idea that this device could be used for "musical telegraphy", multiple telegraphy or perhaps even the transmission of sound [8].

Although their starting points were far apart, Bell and Gray were rapidly converging on the same breakthrough. But their styles were different. Bell worked in secret, telling as few people about his ideas and progress as possible. Gray worked in the open, corresponding freely with experts and even holding press conferences. In this way Bell learned that one of the best workers in the field, a man who knew far more about electricity than he did, was after the same invention [9]. This spurred him on. He rented two attic rooms above Williams' workshop to test out the instruments Watson was building. By January of 1875 he had a working prototype of his multiple telegraph. In February he traveled to Washington to submit a patent application. He demonstrated it to William Orton, the president of Western Union. Orton hinted darkly that Gray had a better apparatus and that Western Union was so large and powerful it could break any inventor.

Alexander Graham Bell

Bell then went to see Joseph Henry at the Smithsonian. The old man of electricity was interested and asked Bell to come back the next day and demonstrate his multiple telegraph. The demonstration went well and Henry gave Bell encouragement. But Bell complained that he lacked knowledge in electrical engineering. Henry's laconic answer: "get it!" [4].

Bell returned to Boston. More work needed to be done. The instrument was unreliable and the vibrating reeds had a tendency to interfere with each other. It was pure drudgery. Bell and Watson spent long hours tuning and re-tuning reeds and making minor modifications. The project was becoming a bit discouraging. During one of those long sessions Bell revealed to Watson that he had another idea: he thought it might be possible to transmit speech using his multiple telegraph. He knew from reading Helmholtz's book that the ear separates the mixture of frequencies into pure tones. His multiple telegraph, he figured, could do the same thing. He could separate the frequencies with a string of reeds, combine the pure tones and send them over a single wire and then reproduce them with an identical string of receiving reeds [4].

Thomas Watson

June 2. 1875 was a hot day in Boston and it was unbearable in the attic above Williams' workshop. As it had done for the past four months, the multiple telegraph was giving trouble. Watson was in one room with the transmitter, Bell in the other with the receiver. Bell was in the process of tuning the reeds on the receiver, holding it close to his ear. The contact on the transmitter was stuck and Watson plucked his reed. Bell, with his trained, sensitive ear, could hear the sound of the reed, even though there was no battery in the circuit [4].

Bell rushed over to Watson and asked him what he had done. Watson explained and, excitedly, they tried it again. Watson screwed down the contacts on the other reeds and plucked them too. Bell could hear all the various pitches faintly in his one receiving reed.

Suddenly it came all together for Bell. Here was the way to transmit sound. The transmitting reeds were part of a magnetic path. The small residual magnetism in the iron was generating a current in the coil. With a permanent magnet instead of the soft iron core and a membrane attached to the reed, he could produce a current modulated by the sound waves of speech [1].

But he realized something else. One reed, damped by pressing it to the ear, reproduced the sounds of all the reeds. It was not necessary to use a string of reeds to transmit sound; a single device, if properly designed, could do the job. Bell sketched such a device for Watson. Watson studied it on the train to Salem and started building it the next day. A day later they tried it. Watson was on the receiving end and he thought he heard Bell's voice. But when they switched positions, Bell heard nothing. They fiddled with it and tried it again a few days later, with the same disappointing results [4].

Hubbard urged Bell to complete the multiple telegraph and not to get sidetracked by a flimsy idea for a different product. Discouraged, Bell dropped all work on the telephone for eight months.

But on February 14. **1876** he submitted a patent application on the telephone, even though he did not have a working prototype. In the application he added an idea which had occurred to him a few weeks before but which he hadn't yet tried. It was an idea for a different transmitter: a stylus attached to a membrane, dipping into a weak solution of acid. The vibration of the membrane modulated the depth of the stylus in the acid, thus modulating its electrical resistance. With a battery in the circuit the changing resistance would modulate the current in the wire and move a reed at the far end [4].

The same day, a few hours later, Elisha Gray filed a caveat with the patent office, describing the identical idea [4]. Bell's patent application was allowed on March 3 and issued on March 10 (the patent office was much faster in those days). But Bell still had no working model.

In the evening of March 10 Bell and Watson started to test the first transmitter (i.e. microphone), which used the variable resistance principle described in the patent application. As a receiver they used one of the tuned reeds. Pressing it against his ear to dampen the reed, Watson suddenly heard Bell say: "Mr. Watson, come here, I want to see you". Watson ran to the next room and told Bell what he had heard. Excitedly they switched positions several times. There was no doubt; the telephone had finally begun to speak [4].

To be sure, it was a very poor telephone. When it was demonstrated to Hubbard three days later, Hubbard wasn't certain he actually heard anything. But Bell and Watson worked at it. A few weeks later, according to Watson, the transmission had improved to the point "where the message didn't have to be repeated more than half a dozen times" [4].

The microphone was a problem. The acid spilled easily, the point in it oxidized and it was difficult to adjust. Bell went back to the magnetic transmitter and finally got it working, too. And instead of a simple reed he had Watson built a receiver, an electromagnet inside an iron cylinder with a thin iron membrane.

The year 1876 was the hundredth anniversary of the United States and a Centennial Exhibition was organized in Philadelphia. Hubbard urged Bell to exhibit his multiple telegraph and telephone. Bell took his devices to Philadelphia and set them up in the Massachusetts exhibit. On Sunday, June 25, 1876 the exhibition was closed to the general public. The guest of honor, Emperor Dom Pedro of Brazil, toured the exhibits, together with his entourage and the official judges, among them Sir William Thomson (later Lord Kelvin) [4].

The group spent a long time at the booth of Western Union, where the main attraction was Elisha Gray's multiple telegraph. Bell stood nervously by while Gray explained and demonstrated his invention. Finally the judges and the Emperor moved on. At the Massachusetts exhibit Bell put his own multiple telegraph into operation. It was similar to Gray's, but somewhat simpler.

When Bell started talking about his telephone, Sir William became interested. Bell had strung a wire and positioned a microphone 100 meters (300 ft) away and he walked over to it. Thomson held the receiver to his ear as instructed. Bell said: "Do you understand what I say?" Thomson ran toward Bell arid told him that he had heard "what I say". When it was Dom Pedro's turn to listen, he jumped up and shouted: "I hear, I hear." Elisha Gray listened and was stunned by the ghostly, ringing voice coming out of the iron cylinder [4].

In September, Hubbard offered Watson one-tenth of the patent interests if he would join the Bell group. Watson hesitated. He had a steady job, was making $ 3.00 per day and was in line for a promotion to foreman. Hubbard's offer sounded risky to him. Finally they compromised; Watson agreed to a contract for half his time, but still received a one-tenth interest [4].

During the fall, Bell and Watson started to improve the telephone's performance by methodically varying every part. They tried membrane sizes from one meter (3 feet) in diameter down to a few millimeters, ending up with an optimum roughly the size of today's telephone receiver. In October they achieved the first two-way conversation over a private telegraph line three kilometers long between Boston and Cambridge. One electromagnetic transmitter was used at either end, each acting as microphone and receiver [4].

Hubbard, in the meantime, had gone to see William Orton of Western Union. He offered to sell all rights to the telephone for $ 100,000. Orton turned him down flat [4].

Bell, whose agreement with Sanders and Hubbard only provided for the payment of expenses, not for a salary, needed money badly. To devote time to the telephone, he had almost totally abandoned his private practice. In the hope of providing an income he started giving lectures on his invention at 50 cents per person. The idea turned out to be a huge success, drawing as many as 2000 people for each lecture. The microphone was usually located a few miles away, connected through a telegraph line. Bell would explain the technical details and then Watson would talk and sing to the hushed audience. Only the few people nearest the receiver could understand anything, the rest merely heard a faint noise, despite the fact that Watson had learned to talk and sing very loud, cupping his mouth into the microphone [4].

Hubbard had handbills printed, announcing that telephone installations could now be rented (the subscriber was expected to string his own wire) [7]. The Bell Telephone Company was formed and by the end of July some 600 telephones were in operation, all point-to-point connections. The user spoke into an opening in a wooden box (loudly and slowly) and then quickly pressed his ear against the opening to listen. One of the biggest problems the telephone initially had was that people were caught by stage fright, unable to utter a word when their turn came to speak [3].

In July, Bell married Hubbard's daughter Mabel and in August went with his bride to Great Britain to secure European patents and sell licenses. Through William Thomson's enthusiasm the telephone had been popularized in Britain and Bell was an instant celebrity. He was invited to give lectures to scientific gatherings. One such gathering of the Society of Telegraph Engineers in October 1877 gives an example of the state of sound quality. One of the participants reported that he had tried the telephone a few days before, with a friend at the other end. He said: "Hi diddle diddle - follow up that", and thought he heard the answer: "the cat and the fiddle". When the friend was asked the next day if he had understood what had been said he answered: "No, I asked him to repeat" [4].

On the formation of the Bell Telephone Company, Bell had been named Electrician (i.e. director of research and development) and Watson was put in charge of manufacturing. But Watson in fact filled both jobs. The initial wooden boxes contained no bell; shouting into the mouthpiece or tapping the membrane with a pencil obtained the attention of the party at the other end. Watson developed a ringer, continued to work on the quality of the sound and found himself in a hectic daily schedule solving manufacturing problems [4].

While Bell stayed in Europe (for an entire year), trouble was brewing for him and his associates in America. In December of 1877 Western Union organized the American Speaking Telephone Company and acquired the patent rights to telephone inventions from Elisha Gray and Amos Dolbear, a physics professor at Tufts College. Western Union was a rival to reckon with. It was the largest company in the world, with a capital of $ 40 million and a net income of $ 3 million per year. It was a national organization and had a huge network of wires. William Orton had made up his mind to enter the telephone business and he expected to wipe out the puny Bell Telephone Company [3].

As sales increased and Western Union's competition began to become formidable, Hubbard decided to create the position of general manager and hired Theodore Vail, the superintendent of the Railway Mail Service. Vail left his secure job and took a pay cut to become the $ 3,500 a year General Manager of a company which was badly under-financed, had now an inferior product and was being squeezed out by the most powerful competitor in the world [11]. He immediately raised more capital and, within six months of his arrival the Bell Telephone Company, was leading again in technology; by acquiring and improving a different microphone design the Bell Telephone Company finally was able to achieve satisfactory clarity and loudness.

Much Ado About *Almost* Nothing

In September of 1878 Vail decided to take the American Speaking Telephone Company to court for patent infringement. Technically it was a lawsuit against a Peter A. Dowd, its agent, but the Bell Telephone Company was clearly locking horns with Western Union.

Western Union relied heavily on the caveat of Elisha Gray. Even though it was filed a few hours after Bell's patent application, a reasonable case could be made that the idea had occurred to Gray first. Gray himself had become convinced in his own mind that he, not Bell, was the real inventor of the telephone. And indeed it looked at first as if Western Union would emerge the winner. But then a letter appeared which drastically changed the picture.

Early in 1877 the Chicago Tribune had run an article, which claimed Elisha Gray as the inventor of the telephone and talked about the "spurious claims of Professor Bell". The reporter had been confused by the fact that Gray originally had called his multiple (tone) telegraph a telephone. Bell wrote to Gray, demanding to set the record straight. Gray apologized to Bell and wrote, "I do not claim even the credit for inventing (the telephone), as I do not believe a mere description of an idea that has never been reduced to practice ... should be dignified with the name invention" [3]. Gray had forgotten the letter. When it was produced in court, Western Union's case collapsed. Its attorney suggested that the case be settled out of court.

Western Union at that time had its own problems. William Norton had died and the company was now in the hands of William H. Vanderbilt, the son of Cornelius Vanderbilt. The infamous Jay Gould was trying to take control away from Vanderbilt, as he indeed later did. He established a rival Atlantic and Pacific Telegraph Company, undercutting Western Union's prices and started to manipulate Western Union's shares on the stock market.

So Vanderbilt was only too glad to settle a bad lawsuit, which took time away from his more serious problems. Western Union admitted that Bell was the inventor of the telephone and agreed to get out of the telephone business. The Bell Telephone Company got the rights to all of Western Union's telephone patents and acquired all of its 56,000 telephones in 55 cities in return for a 20% royalty on rental income for the life of Bell's patent [4].

Elisha Gray regretted his letter bitterly for the rest of his life. Ironically the rights to his telephone inventions were now in the hands of the Bell Telephone Company. In 1885 he reopened his case, claiming that the patent examiner had disclosed his caveat to Bell and that only then Bell had added Gray's microphone design to his patent disclosure. The case went all the way to the Supreme Court and Gray lost again [9].

Like Gray, Amos Dolbear had been abandoned by Western Electric in the settlement of the Dowd lawsuit, but he was not ready to give up. In 1880 he filed a patent application on a system with a slightly different microphone and receiver and incorporated the Dolbear Electric Company. The Bell Telephone Company sued.

Dolbear made an interesting case. He had been thinking about a telephone as early as 1876, but when a friend of his approached Hubbard he was told that Bell was already ahead of him. Dolbear now claimed that he had been told that Bell had already applied for a patent and that was the reason he, Dolbear, did not file a patent application himself.

Dolbear's case looked weak, so he (or perhaps his attorney) decided to prove that Bell's patents were in fact worthless because he had been anticipated by Reis. A copy was made of Reis' apparatus and brought into court but, to the great amusement of judge and jury, the experts and professors could not make it work beyond a few undecipherable squeaks. Dolbear lost and went out of business, but he never gave up his claim of being the inventor of the telephone. He would display a box to his freshman class and say, "That is the first telephone; I invented it" [3].

But the legal problems of the Bell Telephone Company were not over; in fact they were just beginning. An army of inventors and charlatans suddenly filed telephone patents. In 19th-century American history one finds periodic booms, manias for getting-rich-quick schemes. There had been the gold, silver, oil and Florida land booms; now it was the telephone boom. For quick operators it was a way to organize a local telephone company based on a sham patent and sell as much stock as possible before the Bell Telephone Company sued for infringement. Over the next few years there were an incredible 600 lawsuits, every single one of them decided in favor of Bell. Two of them merit special attention.

The Pan Electric Company was organized in 1883 with paper assets of $ 5 million, consisting exclusively of a sham telephone patent (the actual expenditure for the organization of the company was $ 4.50 to have a state seal affixed). Large blocks of shares were distributed as gifts to politically influential people, including a 10 percent ownership to Augustus H. Garland, a former governor of Arkansas. Then the company sold stock to the public. So far the promoters were ahead of the game, but there came a stroke of luck. In 1884 Grover Cleveland was elected President of the United States and he named as Attorney General none other than Garland [4]. Garland, through his Solicitor General, filed suit for annulment of the Bell patents, charging that they had been obtained by fraud. As far as the promoters of the Pan Electric Company were concerned it was not necessary for the government to win the case; they could pocket the investors' money as long as the Bell patents were tied up in court.

But the bubble burst. In 1885 the New York Tribune exposed the stockholdings of the Pan Electric Company. Garland, claiming that he was out of town when the lawsuit was filed, was whitewashed by a friendly congressional committee and not fired by Cleveland. Incredibly the case went on for another 11 years, though with little zest. It was finally abandoned in 1896, two years after the patents had expired [4].

The second case is that of Antonio Meucci. Meucci had emigrated from Italy to Cuba and there started a factory to electroplate swords and belt buckles. He then moved to New York, operated a candle factory on Staten Island and went broke. His next venture was a brewery, which also went bankrupt. From this point on he devoted his time to inventing, working on such things as a coffee filter, the petrification of the human body, a piano with glass keys and a battery to drive a locomotive. He was granted a few patents, but was singularly unsuccessful in making money on his inventions. His English was very poor; he had to use an interpreter when dealing with the business community [12].

In 1871, more than four years before Bell, Meucci filed a caveat for a telephone. It described a telephone alright, but it was a bit mysterious. Two people were shown talking to each other over a wire, each of them electrically insulated from the ground.

Apparently Meucci had discovered his "telettrofono" back in Cuba around 1849 when a man in his employment complained of being sick and Meucci attempted to cure him with electricity. A piece of copper, connected to a wire, was placed in the man's mouth and when Meucci switched on the battery in an adjoining room he heard the man's cry over the wire. Taking up the idea on Staten Island 20 years later he started experimenting, formed a partnership with three of his Italian acquaintances and had a lawyer in New York draw up a caveat. The lawyer told him that he didn't think his invention was quite ready for a patent application and that he would like to come out to Staten Island and see Meucci's apparatus. Meucci never invited him. He renewed the caveat in 1872 and 1873, but then let it lapse [12].

Meucci would have dropped out of memory as one of thousands of unsuccessful inventors had it not been for Seth R. Beckwith. In 1883 Beckwith organized the Overland Telephone Company of New York, based on a dubious telephone patent he had acquired. He sold stock in the company, started to install telephones and was promptly sued by the Bell Telephone Company. Beckwith did not wait for the outcome of the lawsuit but quietly severed all connections with Overland [10].

At about the same time a similar company was formed in New York, the Globe Telephone Company. It acquired the telephone patent claims resulting from the caveat of Antonio Meucci, but never engaged in any commercial activity. It is not clear whether Beckwith had anything to do with the formation of Globe, but in 1885 he became a stockholder and made Globe's office his base of operation.

The 77-year-old Meucci was perfect for Beckwith's plans. His caveat antedated Bell's invention by four years and he could claim that he had a telephone in operation as much as 27 years before Bell. Shortly thereafter the Bell Telephone Company sued Globe for patent infringement.

The case was in court for two years. In the middle of the proceedings Beckwith disengaged himself from Globe and started yet another company, the Meucci Telephone Company of New Jersey, also using Meucci's patent claims. He was sued again and the Bell Telephone Company won both cases [10]. The court proceedings brought Meucci into the limelight. As depicted by the defense, he was the underdog, the poor Italian immigrant whose rightful claims of priority were cruelly suppressed by big business.

What did Meucci invent? In court he was unable to show the apparatus of 1871; his wife had sold it to a scrap dealer. But an examination of the caveat reveals the story. The two men on insulated stands have only one wire between them; there is no return path, no electric current could possibly flow. The transmission of sound over the wire was entirely acoustic; what Meucci had invented was nothing more than the old tin can telephone.

But the myth of Antonio Meucci as the inventor of the telephone has refused to go away. Even today there is a tablet on the Florence post office building which proclaims, "Antonio Meucci - Inventor of the Telephone - Died in 1889 - In a Foreign Land, Poor - And Defrauded of His Rights" [12].

And in Friedrichdorf, Germany, there is a museum and monument claiming Philip Reis to be the inventor of the telephone.

Hubbard's decision to rent telephones rather than to sell them turned out to be a profitable one in the long run. For the infant Bell Telephone Company, however, it was nearly disastrous. There was little money coming in, yet the expenditures for equipment and lines were increasing at an alarming rate. Hubbard raised money, but in the process control of the company passed into the hands of a group of Boston financiers. By March 1879 Hubbard was no longer president [4].

The new capital had been raised at a price of a few dollars per share. As the negotiations for the settlement with Western Union proceeded, the stock value rose. By the end of March the price was $ 50; by September it had jumped to $ 300 and on November hit reached a high of just under $ 1000.

But as profits and dividends rose, so did the problems. Almost all wiring was overhead. During each snowstorm an alarming percentage of the wires collapsed. Noise in the lines also became a serious problem. A single wire was used for each subscriber, with the ground as a common return for all connections. Not only could one hear several conversations in the background, but as electrical power lines and streetcars began to appear, the buzzing, clicking and whirring became seriously annoying. Vail pleaded for the allocation of money to change the system over to two wires for each line (as it is now) and start putting the major lines underground, but the financiers were not interested. Frustrated, Vail resigned in 1887 [11].

The Bell Telephone Company continued on its path of serving only urban areas. In 1885 the American Telephone and Telegraph Company was formed as a subsidiary in New York to specialize in long-distance lines.

In ignoring the rural areas the Bell Telephone Companies had left the door open for competition. In 1893 Bell's patent expired and by 1900 there were some 6000 independent telephone companies in the Unites States. They started to string their lines into the cities, but AT&T refused to interconnect. As a result, some cities had up to three separate telephone systems. A business that depended on the telephone needed two or three sets of telephones and it was impossible to place a call from a remote city into a rural area [7].

Because of the unexpected competition and the ever-increasing demand for capital, AT&T suddenly found itself on the brink of financial disaster and by 1907 control of the company went to J.P. Morgan. Morgan quickly hired Vail, now 62, as president. Vail changed the direction of the company drastically and applied some heretical concepts: maximizing profits, he said, is not necessarily the primary objective; there must be a proper balance between profits and public service. He started disclosing the company's affairs to the public, maintaining "if we don't tell the truth about ourselves, someone else will". He listened to the problems of managers in small groups and hired people by ability, not breeding and set up a central research laboratory in New York [3].

But both Vail and J.P. Morgan were dead-set against competition. They started buying up independent telephone companies by the dozen, often by cutting off credit through Morgan's vast banking connections, thus starving their targets into submission. In 1910 AT&T even took over Western Union.

With the ruthless takeover of independent telephone companies AT&T had made enemies. A movement grew to enforce the Sherman Antitrust Law against AT&T and, when J.P. Morgan died in 1913, Vail volunteered to negotiate with the government. He was not against government regulation in moderation but he felt strongly that the telephone should be a monopoly. AT&T signed a consent of monopoly with the government under which it agreed to dispose of its Western Union Stock, to make no more acquisitions of independent telephone companies without government approval and to interconnect with other phone companies [3].

AT&T's monopoly status lasted until 1983 when the government forced the breakup of A.T.&T and its 10 regional phone companies. Before long the 11 independent companies started merging and by 2006 there were only three, the largest of which was A.T.&T again.

Alexander Graham Bell left the Bell Telephone Company in 1879. Settling in Washington, he bought a mansion, which occupied an entire city block and acquired an estate on Canada's Cape Breton Island a few years later. By 1884 his Bell Telephone Company stock had appreciated so much that he was a millionaire [4].

He continued doing research, mostly alone or in small groups, working on the transmission of sound over light beams (which he felt was his greatest invention), an improvement on the phonograph, sonic methods for determining ocean depths (now called sonar), aeronautics, genetics and hydrofoil boats [4].

But his main interest remained the deaf. When asked his profession he would always answer: "teacher of the deaf". He championed lip reading, invented an "audiometer" to test the hearing and contributed a significant portion of his fortune to causes for the deaf.

When he died of diabetes on August 2. 1922, the entire telephone service in America was silenced for one minute.

And what happened to Thomas Watson? He resigned his position with the Bell Telephone Company in 1881 and went on a yearlong trip to Europe. He got married and bought himself a farm in East Braintree, south of Boston. But he found that farming didn't agree with him, so he converted a barn into a machine shop and before long started to develop a rotary steam engine. He never got this engine working well but his next project was a success: a steam engine for yachts and tugboats, which soon proved to be the best of its kind [15].

During this period he educated himself, taking courses in geology and, inspired by Bell, in elocution. Orders for his steam engine poured in and within a short time he was running a full-fledged business. He expanded into the manufacturing of printing presses, shoe machinery and boat hulls. Then he opened a shipyard, eventually building entire destroyers and battle cruisers, employing 4000 people.

But by 1903 he was in over his head. In a downturn he lost control of his company and his entire investment. All he had left was a small trust fund [14]. Characteristically, the happy-go-lucky Watson didn't stay depressed for very long and started to pursue his other interests, geology and elocution. Elocution led to public speaking and then to acting. At age 56 he spent a year in Great Britain as a Shakespearean actor. He died at age 80 in 1934, not rich, but happy.

6. Tesla

The world's greatest inventor
and how Nikola Tesla trumped him
and then went over the edge.

With the discovery of induction it became possible to convert mechanical energy into electricity. So we would now expect an explosion of applications, leading in a straight line to today's huge power plants. But in reality very little happened for a long time.

What you get with induction is quite different from a battery. A battery delivers a steady current (for as long as it lasts); when you reverse the connection the current flows the other way. Simple and easy to understand.

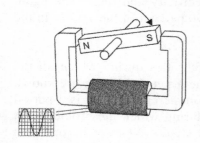

The electricity produced by induction is radically different: it has to change constantly. Picture a strong magnet affixed to a water wheel or a steam engine, rotating inside a U-shaped iron arrangement with a coil wrapped around it. As the north pole of the magnet approaches the top end of the iron and the south pole the bottom, a voltage builds up at the terminals of the coil. As the magnet poles move past the iron the voltage decreases, goes through zero and then becomes negative as the south pole approaches to top end of the iron U and the north-pole the bottom. What you get is a sine wave. It's a beautiful waveform, a mathematician's delight, but it's not what you want.

The natural waveform produced by a generator is AC, not DC

To the people of 1831 this was very confusing. What could one possibly do with a voltage which changed its polarity several times a second (and, consequently, a current which constantly changed its *direction*)?

A few months after the discovery of electromagnetic induction a Frenchman by the unlikely name of Hippolyte Pixii found a solution after a fashion. Pixii built a generator, a small, hand-cranked machine in which a U-shaped permanent magnet twirled around, its poles facing an equally U-shaped iron core with a coil wound on it. The product of the coil was - naturally - alternating current, but he connected the wires to an arrangements of contacts which opened and closed by the rotation of the magnet so that the polarity of the coil was reversed every half-turn. In this way the voltage that appeared at the terminals of the coil always had the same polarity. It wasn't a steady voltage by any means, more like a series of pulses, but at least the current flowed in one direction only.

Hippolyte Pixii received patents on his invention, but they didn't make him rich. There was no economic application for such a generator yet, it produced very little power. A number of other experimenters improved on Pixii's generator, building larger machines powered from water wheels or steam engines, but they couldn't compete with mechanical systems. In machine shops and factories overhead shafts and wheels supplied this power through belts to machines. Though this system was cumbersome, ugly and often dangerous, it was efficient. It made little sense to convert mechanical power into electricity and then back again into mechanical rotation, only to lose some 90 percent of the energy in the process.

Thus we now enter a period of some 50 years during which all the fundamental principles of electric power generation and light were known but couldn't be utilized in any significant commercial way. It is a period without any startling discoveries, a period during which the efficiency of generators slowly increased, step by minute step, driven not by one or a few glamorous inventors, but by a large number of lesser-known tinkers and experimenters.

There was also another potential application: the electric light. Experimenters had noticed for years that a current could heat up a thin wire. As they increased the current the wire would produce a red glow which became brighter and brighter. But just as it started to shine brightly it burned up.

Enter Thomas Alva Edison, the world's most prolific inventor.

Edison was born in 1847 in Milan, Ohio. His mother called him Alva; to his friends he was Al. He did poorly in school; his teacher thought that he might be retarded. So his mother took him out of school and taught him herself. He was unruly, couldn't sit still and his mind wandered. He learned primarily by discovering things himself. Spelling was a bother to him; all his life his grammar and syntax were appalling.

He was puny as a child with a head that somehow seemed too large. His coordination was poor and he never became interested in sports. He had few friends and spent much of his time reading [7]. His father disapproved of him, regularly beat him and thought he was stupid.

At age 12 he set up a chemistry lab in the basement, staying up long into the night. At 15 he added electricity to his nighttime activity and built himself a telegraph. At 16 he left school and became an itinerant telegraph operator, moving to a succession of towns: Port Huron, Stratford, Adrian, Toledo, Fort Wayne, Indianapolis, Cincinnati, Memphis, Louisville, Montreal, Boston.

In January of 1869, not quite 22 years old, he quit his job in Boston and announced that he was going to be an independent inventor. His objective was to make money, large amounts of it, and as quickly as possible. Making money was the main driving force in his life [7].

It was a bad start. He found a backer for $ 100 to develop and market an electrical vote-recording machine. He tried to sell it to the legislature but had no luck. His second venture didn't fare any better and his backers withdrew. With his last few dollars he went to New York. And here in New York Edison finally found some success. The stock and gold tickers had an annoying problem: if there was any disturbance in the system, an operator had to go to each receiver and reset it, so that the entire system would start up in unison. Edison felt sure he could come up with a system, which would reset automatically [7]. He formed a company with two partners and developed a printing telegraph on a shoestring budget. It was Edison's first working instrument. A few months later the three partners sold their enterprise to the Gold & Stock Company (later to be absorbed by Western Union) for $ 15,000. Edison's share was one-third [7].

He was now working at a furious pace, determined to become rich quickly. He made a deal to develop half a dozen devices for Western Union and went into partnership with William Unger, a machine-shop operator in Newark. The firm of Edison and Unger was to manufacture the devices for Western Union.

Then he started another company with George Harrington, a former secretary of the treasury. The American Telegraph Works, in which Edison had a one-third interest, was to produce an automatic telegraph repeater in direct competition with Western Union. At the same time he worked both for and against Western Union [7].

Before the product was even developed he rented a huge building in Jersey City and bought $ 30,000 worth of equipment in the first week, simply sending the bills to Harrington. Harrington, aghast by the extravagance, wondered if Edison thought he was still secretary of the treasury.

Edison hired a team of talented men who, like him, were willing to work long hours. Among them was Charles Batchelor, an Englishman, and John Kruesi, a Swiss machinist. They complemented each other perfectly. Edison, who had poor manual dexterity, came up with the ideas, Batchelor translated them into drawings and Kruesi made the models. His men were highly motivated: they were all going to be rich with him.

On Christmas day 1871 Edison married Mary Stillwell. The next morning he was back at work with his boys. When his first daughter was born he wasn't there; he was too busy with his work [7].

It was an exciting time for Edison and his men. They were working on dozens of new inventions, all at the forefront of technology. But trouble was brewing rapidly. The automatic repeater had problems in production. Edison had hired an incompetent manufacturing manager and the American Telegraph Works was in shambles. Edison's accounting system was murky; it consisted mainly of getting more income when the sheriff was at the door. Most bills were paid in installments; some were never paid. He got into a dispute with Harrington and walked out, taking his team with him. Unger, dissatisfied with Edison's erratic management and draining of profits through heavy research spending, wanted to get out of the partnership [7].

Things were now getting complicated. Edison was making deals left and right and no longer knew what he had signed. In December of 1875 he bought several acres near the railroad station in Menlo Park, New Jersey, built a long, two-story building and moved his crew into it, together with $ 40,000 worth of equipment and supplies. He wanted to get away from distractions and concentrate on research and development. What he had in mind was an "invention factory", producing a minor invention every 10 days and a "big trick" every six months. But it was mostly wishful thinking; he was constantly dissipating his time on manufacturing problems. While setting up Menlo Park he also incorporated the American Novelty Company, which was to manufacture an artificial perfumed rose for buttonholes, a mechanical flying bird, and "Edison's Perpetual Segar", a refillable cigar that was to drive pipes out of existence. All three products failed [7].

It was probably in the fall of 1877 that he first thought of the possibility of recording the human voice. He was working on an automatic telegraph repeater, which had the dots and dashes of the Morse code indented on paper. The idea was to open and close a sensitive electrical contact with the indentations. The strip of paper could be pulled through the machine at different speeds, so that the message could be run over the line at high speed, recorded at the other end and played back at a speed slow enough for the operator to write down the message [7].

When he ran the strip at a high speed he was puzzled by the sound it made, a sound similar to a high-pitched human voice. He made a sketch for Kruesi: a hand-cranked cylinder with a spiral groove on it and a tin foil placed over the cylinder. Along one side of the cylinder ran the recording device, driven by a screw. It had a stylus in the center of a membrane so that its vibrations pushed the soft foil into the groove on the cylinder. With a separate stylus and membrane at the other side of the cylinder he could then convert the impressions in the tinfoil back into sound. To both Kruesi's and Edison's astonishment it worked the first time. There was the human voice, faint and scratchy, but without doubt the human voice, coming from a machine.

Edison took his phonograph to New York, to the editor of Scientific American. He created such a sensation that the editor was getting worried the floor would collapse from all the people who wanted to hear the new miracle. The Scientific American ran a story in the December 22 issue, spreading the word all over the nation [7].

In January 1877 Edison raised money and formed the Edison Speaking Phonograph Company. He received $ 10,000 in cash and a 20% royalty. For a short six months the Edison Speaking Phonograph Company was very profitable. Demonstration teams toured the cities, playing Edison's speaking machine for an admission fee and collecting as much as $ 1800 per week. But then the novelty suddenly waned. Nobody wanted to buy the machines; the quality of the reproduced voice was simply too poor. The voice on the machine could only be understood if the people had been present during the recording and expected to hear what they thought they heard. If they listened to a prerecorded foil, they heard nothing but a badly garbled murmur. In addition, the foil was so soft that it could only be played back half a dozen times. Edison lost interest in the phonograph and for eight years did no work on it. He became absorbed by the electric light.

Thomas Edison

Edison, now 30 years old, had a staff of 40, including two chemists and a physicist, and his equipment and machine shop were probably the best in the world. Although Edison rarely wrote articles himself, he carefully read the scientific literature. He ran across an account of experiments with light bulbs made by a Joseph Swan and became intrigued. He began an investigation of his own by heating up small strips of carbonized material in open air, an investigation that Swan had put aside 30 years earlier. Edison prematurely concluded that carbon was not a suitable material for a light bulb.

Grosvenor P. Lowrey, the general counsel for Western Union, proposed that a new company be formed to exploit the electric light. Edison agreed and Lowrey started to negotiate with potential investors.

In October 1878, while negotiations were going on, Edison made announcements to the press, hinting that he had "solved the problems with the electric light" [7, 13]. In fact he had accomplished nothing of the sort, he was still far behind Swan and on the wrong track: he was heating platinum wire in open air. But his bragging to the press helped him to negotiate a good deal. In November 1878 the agreement was signed for the financing of the Edison Electric Light Company. Edison received five-sixths of the shares of the corporation, a $ 55,000 cash payment and a 5-cent royalty on every light bulb to be produced. When he received the first $ 30,000 of the cash payment he didn't bother to pay off badly overdue debts, but immediately started construction of an office and library building and expanded the machine shop.

Although Edison hadn't yet the faintest idea how to do it, he knew exactly what he was after. He had studied the business of distributing gas for home lighting. Each individual flame could be turned on and off without affecting the other flames. The only practical electrical light then in existence, the arc lamp, was quite different. Because of their low resistance, arc lamps had to be connected in *series*, i.e. the current would flow in one path through a string of lamps. If one lamp went out (which happened quite often) they all went out.

Intuitively Edison realized that a truly useful electric light needed *parallel* connection: two wires from the generator to which each lamp was connected through a switch. But this meant that the filament of each electric lamp would have to be of high resistance. The task was in fact more difficult than anyone had imagined. Not only was a material required which would not fall apart at extremely high temperature, but it had to be very thin as well.

In December of 1878 Francis Upton joined Edison's team. Upton was from Boston and had a solid education; Edison took one look at him and nicknamed him "Culture". Culture's first assignment was to conduct a thorough literature search on all previous work on the electric light.

Having formed a company and announced that he had the solution, Edison now started to work on the light bulb in earnest. He tried metals with high melting temperatures: platinum, iridium and tungsten, sealing them with an old vacuum pump Upton borrowed from Princeton. All the metals tried behaved the same way: they produced a bright light and then quickly melted. Edison tried several schemes to regulate the temperature of the wire, with limited success. In March 1879 the shareholders demanded to see the light bulb in operation and Edison put on a demonstration. It was a complete fiasco; the lamps flickered and burned out almost immediately.

Edison tried ever more exotic metal combinations to obtain the highest possible melting point. He knew there were two problems to be conquered: to find a material that remained strong at high temperature, and to remove more of the air in the bulb to keep the filament from oxidizing [7].

In August 1879 Ludwig Boehm joined the laboratory crew. He was a glass blower from Germany and introduced Edison and Upton to the latest techniques in vacuum pumps. By September 26. they had a platinum-iridium light bulb, which lasted 13 hours and 38 minutes [7].

In the meantime Edison had been thinking about a generator for his light bulb. He was dissatisfied with the efficiency of available generators (the best one had an efficiency of only about 55 percent), so he decided to build his own. It turned out to work very poorly and he gave the job to Upton who, with his knowledge of mathematics, made the generator work with an efficiency of about 80 percent [7].

But things were getting a bit desperate for Edison. He had long spent the $ 55,000 and, despite his newly acquired expertise in vacuum technology, was nowhere near a practical light bulb. No matter how exotic (and expensive) a metal he used, all the filaments tried so far had a nasty tendency toward self-destruction. A slight over-voltage from the generator and the light bulb was gone.

Now Edison went back to carbon, probably because he read a just-published account of Swan's progress. He experimented with a thin carbon rod in vacuum and then tried to make carbon spirals, but failed. Next Edison carbonized a thread and, after a few attempts, was able to mount it inside a glass bulb, pump the air out and light it for a few minutes [7].

Swan was still ahead at this point, but Swan was satisfied with using low-resistance carbon rods for his experiments, while Edison was unshakably convinced that a commercially successful light bulb needed to have a high resistance. He figured that he could achieve 100 ohms of resistance and that a light bulb needed 100 watts of power to be bright enough. Thus the voltage across the bulb would be 100 Volts and the current through it 1 Ampere. He allowed for a 10 percent voltage loss in the wiring, thus he set the generator voltage at 110 volts. It was this consideration of Edison's that ultimately led to the standard household voltage of 110 volts in the United States.

Edison now furiously tried every conceivable material which he could convert into carbon by baking it at a high temperature: fish line, cotton, cardboard, tar, celluloid, coconut hair, wood shavings, cork, and even a hair from Boehm's red beard. Although all these materials ended up as carbon once they were baked, their inside structure and thus strength in carbonized form depended greatly on the microstructure of the original material.

On November 10, 1879 the team managed to make a filament out of cardboard, boiled in sugar and alcohol and then carbonized, which lasted for 16 hours [7]. It had less than one-tenth the light of a modern 100-watt bulb and the glass envelope became black inside long before the filament burned out. But it worked, and for the first few hours Edison's bulb produced a light brighter and more pleasant than a gaslight. Ironically the material and shape of the filament was almost identical to the one Swan had tried 31 years before; the only difference was the better vacuum.

The New York Herald carried a story on Edison's light bulb on December 21, **1879**. Within hours there were so many visitors to Menlo Park that the railroad had to schedule extra trains. The shares of the Edison Electric Light Company soared to $ 3,500 bid, $ 5,000 asked. At age 33 Edison now was worth close to $ 1 million on paper.

In January 1880 one bulb lasted for 40 hours; in February 550 hours was reached. In July Edison absentmindedly plucked on a bamboo hand-fan used to dry liquids in the laboratory. It suddenly hit him that carbonized bamboo fibers might be a suitable material for a filament. He tried one - it was test number 1253 - and it turned out to be the best material yet. Never one for half-measures, he conducted a world-wide search for bamboo, testing over 6000 varieties, and found that the best bamboo for this purpose grew in Japan. For the next 14 years millions of

Much Ado About *Almost* Nothing

Edison light bulbs were made from the bamboo of a single grove adjacent to a Shinto shrine in Kyoto [15].

In April 1882 Edison moved production of the light bulb to a plant in Harrison, New Jersey. By summer the plant was producing 1200 light bulbs per day. But Edison's vision extended far beyond the production of light bulbs. He wanted to supply the entire system, generators, wiring, switches and light bulbs and he decided to start by lighting up New York City.

He drew up plans for the first phase of electrifying New York, a system which was to supply one square mile, at a construction cost of $ 160,000. This work was to be the last hurrah for Menlo Park. The machinery and many of the people were transferred to New York for the huge project. Edison moved to an office in town and stayed there most of the time.

In a mad rush Edison started a jumble of new companies, some to install power plants, others to manufacture the various accessories, such as switches, fuses and meters. The Edison Lamp Company supplied the light bulbs. There were now also several Edison power companies in Europe.

Enter Nikola Tesla. He is almost forgotten now, but it was Tesla who solved the conundrum of the naturally occurring sine wave and gave us the power system we have today.

Nikola Tesla was born in 1856 in what is now Croatia, but was then part of the Austro-Hungarian Empire. He was a strange boy with few friends. At school he was brilliant in languages and superb in mathematics; he irritated the teacher and the rest of the class by having the answer before the teacher even finished stating the problem.

He often experienced images in his mind, sometimes preceded by a flash of light. The images were so vivid that he could touch them with his hand and was unable to tell whether they were real or imagined. He went on mental journeys, saw new places and countries, lived there, met new people and made new friends. These journeys were as real to him as if he had actually traveled there. At 17 this mental imagery transformed itself gradually to invention: he was able to picture novel machinery in great detail and in full operation [27].

At age 19 he entered a polytechnic school in Graz, Austria. At first he studied intensely, getting up at 3 a.m. and working until 11 p.m. seven days a week. One teacher became worried that Tesla was damaging his health and warned his parents.

During Tesla's second year at Graz the school received a small electric generator for classroom demonstrations. As all generators of that time it had a ring of contacts on the rotating part, which switched the polarity back and forth so that the output was converted to DC. Tesla observed the sparking, noisy and smelly contacts and proclaimed that they were not necessary. He got into an argument with the teacher, who proceeded to demonstrate that without these contacts the generator could not possibly work.

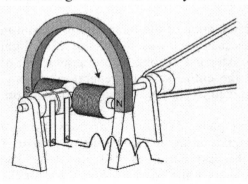

To avoid AC, early electrical generators used contacts to "rectify" the output. Although it didn't produce a steady current, at least it flowed in one direction only.

In his third year Tesla became bored with school, started gambling, got poor grades and flunked out.

At 24 he started again at the University of Prague, but left after only one term, without getting a degree. He then moved to Budapest and worked as a draftsman at the Central Telegraph Office. He had been thinking about the generator in Graz and how the contacts could be eliminated. The problem haunted him and started working long hours again to come up with a system that was purely AC. Deprived of sleep, he had a nervous breakdown.

In early **1882**, at age 25, the solution came to him in a brief flash: there needed to be not just one winding on the generator, but several (he eventually settled on three). As the rotor turned the first winding would reach its peak, followed by the second and the third and then the first one again. In

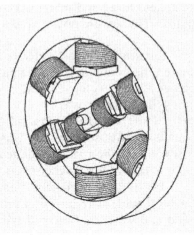

In his revolutionary AC system Tesla used multiple wires with the voltages staggered in time. The inner part, the rotor, was simply a piece of iron with a coil that became a strong electromagnet with rotation. No contacts were needed.

the motor, which was almost identical to the generator, again three coils were used. If the rotor consisted of a permanent magnet, it would follow the rotating magnetic field of the windings. With two "phases" (i.e. coils) the direction of the rotor was uncertain, it could turn either way; with three the direction was assured.

But he didn't stop there, he added his most important invention: if the rotor was made of soft iron and was wound with its own coil with the ends connected together, the rotating magnetic field produced by the stationary part would induce a voltage in the inside coil, which in turn would create its own magnetism. The rotor thus would start turning in the direction of the rotation. If it turned more slowly than the field, the induced power would be large, gradually decreasing as the mechanical rotation approached the electrical one. Thus such a motor produced a lot of power during start-up, a highly desirable feature. And the rotor needed no contacts [28].

Nikola Tesla

Tesla did all of this in his head; he didn't need to make any drawings. He could operated the system in his mind, detect flaws and optimize performance.

In April 1882 Tesla moved to Paris to work for Edison Continental. He tried to interest the company in his AC ideas, but found no response. In the summer of 1883 his assignment was to fix a power station in Strassburg (now Strasbourg). There had been an unfortunate explosion during the dedication ceremony, with the Kaiser present. Tesla found the problem had been caused by faulty wiring and proceeded to bring the station into shape. And there, in his spare time, he managed to build three small generators demonstrating his AC system. But he still had no luck convincing anyone.

He decided to move to New York to talk to Edison. He arrived there in June of 1884 and found a job at the Edison Machine Works. There are several versions of meetings between Edison and Tesla (all written by Tesla and all different) of which this one is probably the most accurate: he met Edison only once, while the latter was on a holiday at Coney Island with his wife. The meeting had been arranged so that Tesla could talk about his ideas, but Edison got distracted and Tesla never got a chance to make a presentation [25].

Edison was now a big man, a tycoon with dozens of companies and a world-famous inventor. Even if Tesla had been able to make his case, Edison had far too much invested in his DC system. And the two of them never would have had a chance of getting along: Edison had a rumpled appearance (he often slept in his clothes), chewed tobacco and spat it on the floor and used earthy language; Tesla was always neatly dressed, deathly afraid of germs and took great pride in speaking English (and seven other languages) with perfection. Tesla approached problems with a mathematical mind; Edison loathed mathematics.

In early 1885 Tesla found financial backers, left the Edison Machine Works and started the Tesla Electric Light Company [26]. He developed a novel arc light, quarreled with his backers and the company went under. For a year he found only temporary jobs and even had to dig ditches for two dollars a day, but then he found new backers. In April 1885, now 30 years old, he formed the Tesla Electric Company with two partners, filed his AC patents and started on the development of equipment. In May **1888** he presented a paper to the American Institute of Electrical Engineers [28]. It was a stunning, masterfully written paper and a man named George Westinghouse read it.

George Westinghouse was born in 1846, which made him a year older than Edison and 10 years older than Tesla. His family came originally from Germany and had changed the name from Wistinghausen. At 16, during the Civil War, he enlisted in the infantry and ended up as an

engineer in the Navy. His father, a talented inventor, established a shop in Schenectady, New York to manufacture steam engines and agricultural and mill machinery [23].

Westinghouse mustered out of the Navy in 1865 and began to attend college. But he found it hard to pay attention in class, sketching locomotives and steam engines instead. After only three months he left college, worked in his father's shop and started to invent [16].

In 1869, at age 22, he received his first major patent. At that time a brakeman stood between every two railroad cars, turning hand-wheels to brake the train when the engineer sounded a whistle. Braking was thus very uneven and delayed. It was worse on freight trains where the

George Westinghouse

brakemen had to sit on top of the cars in all weather. In railroad crashes the engineers often saw each other but couldn't do anything to stop in time.

Westinghouse had the idea of activating the brakes on all the cars from a single high-pressure reservoir. At first he thought of using steam from the engine, but he realized that steam would condense and freeze in the lines in winter. But then he happened to read a magazine in which the construction of the first Alpine tunnel was described. In the tunnel the jackhammers were powered by compressed air, which also provided fresh air supply for the workers.

Westinghouse conceived that he could use this method for the braking of trains, pumping up a reservoir of compressed air with the power of the engine and then releasing the air into a hose running along the train and connected to pistons on the wheels. And he made his system fail-safe: the air-pressure kept the brakes *away* from the wheels; if for some reason the pressure failed (e.g. when part of the train de-coupled) the brakes would automatically activate.

He moved to Pittsburgh and raised money to form the Westinghouse Air Brake Company. Westinghouse not only had an inventive mind, but also considerable organizational and marketing talent. The Westinghouse Air Brake Company became a solid success.

By 1884 he was looking for new areas of business and became involved in the distribution of natural gas. Playfully he had a drill installed in the backyard of his house - and found a large pocket of natural gas.

As the light bulb started to compete with the gaslight, Westinghouse became interested in electricity. He heard about an exhibition in Turin, Italy, in which AC was transmitted over a distance of 80km (50 miles) using a *transformer* to step the voltage up to 2000 Volts.

A transformer is simply an iron core with two (or more) coils around it. The first coil is connected to the generator. If the voltage is DC, a magnetic field is produced in the iron but nothing happens in the second winding. But with AC the constantly changing magnetic field induces a voltage in the second winding; if you give the second winding more turns than the first, the resulting voltage is larger. Conversely fewer turn results in a smaller voltage. Thus the transformer lets you increase or decrease a voltage at will. For power transmission over long distances a higher voltage means less current is flowing (for the same amount of power), which means thinner wire can be used.

With the instinct of a businessman Westinghouse saw the advantages of the transformer immediately. The biggest problem with Edison's DC system was distance. The voltage was fixed by the requirements of the light bulb and, since there was no equivalent of the transformer in DC, the voltage had to be the same for the entire system. This was acceptable for small systems, a generator that powered only a house or a few houses in the neighborhood, but in a network covering a larger distance and supplying higher currents, the copper conductors had to be large and expensive to prevent voltage loss along the lines.

By 1886 the Westinghouse company had a small AC system working, feeding 3000 volts over a 1.2 kilometer (3/4 mile) line to six transformers distributed in the basements of houses around the town, running at 133 Hz.

That year Edison had 58 DC central stations in operation with a total capacity of 150,000 lamps. But within a year the Westinghouse Electric Company began gaining rapidly on Edison.

Edison became furious. He intuitively disliked AC - to him it was unnecessarily complicated - and he now had to defend an established position as a manufacturer. He started fighting AC with every means at his disposal. He had his people spread stories that even small AC voltages were lethal while a DC voltage of as much as 1000 volts was safe (no truth to that). The very first product his new laboratory at West Orange worked on was an electric chair for Auburn State Prison, using AC (and for good measure a Westinghouse generator). Edison tried to introduce the term "westinghoused" for electrocuted, but fortunately failed.

No matter how furiously and unscrupulously Edison fought against AC, the technical and commercial realities turned customers more and more against him. To make matters worse, a French syndicate in early 1887 cornered a large portion of the world's copper production, driving prices up from 18 cents per kilogram to over 40 cents. Because of their lower voltages and higher currents, DC stations used much more of the costly copper than AC ones [24].

There was another problem with DC power systems: because of their limited range, the generators and their steam engines needed to be in the center of populated areas. Under Edison's scheme New York City, for example, needed 36 separate power plants. As more and more of them were built, complaints grew about the noise and noxious fumes and the constant traffic of wagons bringing coal and hauling away ash.

But AC systems at that time could only power light bulbs; there were no AC motors. Tesla's revolutionary paper described a system that could do both, and neither the generators nor the motors required contacts to the rotor.

Westinghouse met Tesla and within two months of his presentation they struck a deal: Westinghouse offered Tesla's company (of which Tesla owned 4/9) $ 75,000 in cash, plus $ 2.50 per horsepower of motors sold, in return for all of Tesla's AC patents [14].

Tesla moved to Pittsburgh as a consultant. The Westinghouse AC system used 2 wires and 133 cycles per second (now called Hertz, abbreviated Hz); Tesla's 3-phase system needed 3 or 4 wires and he calculated that 133Hz was too high for his motors; he wanted 60 Hertz. The Westinghouse engineers refused to change an established product. After only nine months Tesla quit and moved back to New York. A year later all work on Tesla's AC system stopped at Westinghouse.

But fate takes strange turns. In 1891 a partnership of a German and a Swiss company demonstrated an AC system in Germany. The generator was at Lauffen on the Neckar river and the 210 kilowatts of power were transmitted at 30,000 Volts over a distance of 175km (110 miles) to an exhibition at Frankfurt, using wires only 4 millimeters (less than 3/16") thick.

The head of the project for the German company, Russian-born Michael von Dolivo-Dobrowolsky, claimed he invented the system, but his Swiss partner, C. E. L. Brown, stated that "the 3-phase system as applied at Frankfurt is due to the labors of Mr. Tesla, and will be found clearly specified in his patents" [25].

Jarred by this development the Westinghouse engineers changed their minds and resumed their works on Tesla's approach, using a frequency of 60 Hz just as Tesla had wanted. This became the standard in the U.S., while Europe eventually settled on 50 Hz

Tesla suddenly became famous; he was the man who trumped Edison. At the 1893 Columbian Exposition in Chicago, which was lit by 180,000 light bulbs powered by Tesla's AC system, he was given his own exhibit. Three years later the first large hydropower station went into operation at Niagara Falls, using Tesla's AC system to transmit power to Buffalo, New York.

When Tesla left Westinghouse in 1889 he opened a laboratory in New York. He was rich now and his two partners agreed to leave the entire $ 75,000 received from Westinghouse in the company; (considering inflation this would amount to $ 1.5 Million today). He hired two laboratory assistants and a secretary and started to spend large sums on equipment. His own lifestyle now spelled affluence. He lived in an expensive hotel and had dinner nightly at Delmonico's, where he had a reserved table, which nobody was allowed to use, even if he wasn't there.

Tesla cut a dashing figure in those days. He was 2 meters tall (6'6") and very thin, weighing just 65 kg (142 lbs). He spoke 8 languages and his English was almost accent-free. He always wore a Prince Albert coat and Derby hat, stiff collar, cane and gray suede gloves; the gloves he wore for a week and then threw them away. He wanted to be the best-dressed man in New York and probably was.

But beneath the worldly exterior was a very strange man with a large number of unusual phobias and hang-ups. He had an inordinate fear of germs: he washed his hands constantly, refused to shake hands and in his laboratory he had his own bathroom, which no one else was allowed to use. Handkerchiefs he used only once and then discarded them [21, 26]. At Delmonico's he required a stack of napkins with which he proceeded to wipe the silverware and then dropped them on the floor.

He needed to count steps while walking and any repetitive task needed to be divisible by three. He had to calculate the cubic content of soup plates, coffee cups and pieces of food; otherwise he could not enjoy his meal [35].

He could not touch the hair of other people, would get a fever looking at a peach and a piece of camphor anywhere in the house would give him great discomfort. He had a violent aversion against earrings on women and the sight of pearls would give him a fit.

In his laboratory Tesla began to experiment with higher AC frequencies. He built a series of special generators with a large number of poles, spinning them very fast and reaching a frequency of 25,000 Hertz [22]. When he couldn't push his rotating machines any higher in frequency he turned to what we now might call electronic approaches. Other people had found that an inductor (a coil) and a capacitor connected together could resonate in a fashion similar to a church bell or a tuning fork. If you applied a voltage to such a resonant circuit, the capacitor would store it (briefly) and then deliver a current to the inductor. The inductor would store the current (briefly) and deliver a voltage back to the capacitor. Thus the energy flowed back and forth between inductor and capacitor and the time it took to do that was determines by how many turns the coil had and how large the capacitor plates were.

Tesla took this basic circuit and augmented it considerably. First he added a second winding, with many more turns than the first one, thus creating at the same time a resonant circuit and a transformer. He found that, at the higher frequencies, he needed no iron core; with wires wound on a simple cardboard tube he could produce several thousand volts across the secondary coil.

At first Tesla's resonator only delivered the high voltage briefly, akin to a church bell being struck once, with the sound then dying out. To make the circuit resonate continuously several experimenters had used spark-gaps (no vacuum tubes or transistors yet, which we would use today). A spark-gap has a curious behavior: when the voltage across it is so large that a spark appears, it abruptly becomes a short-circuit. Thus the voltage across the gap collapses and the spark-gap becomes an open-circuit again. Properly applied, a spark-gap is thus able to feed pulses of energy into a resonant circuit at just the right moments and keep the oscillation going.

Tesla improved on this principle greatly, designing ever more elaborate spark-gaps. He used the resulting high frequency, high voltage generators to produce some stunning effects: a shower of sparks, spidery figures inside a phosphorus coated glass sphere, or making his own body and clothing emitted glimmers and a halo of splintered light [21, 26]. He cultivated journalists and the rich and famous, spending money liberally by giving elaborate banquets and afterward inviting the guests to his laboratory for demonstrations.

In May of **1891** he presented another important paper to the American Institute of Electrical Engineers, this time on his high-frequency work [28]. It was again a stunning masterpiece. The following year he was invited to read his paper at the Institute of Electrical Engineers and the Royal Institution in England and at the Societé Internationale des Electriciens and the Societé de Physique in France. Back in the U.S he topped it all by giving a lecture and demonstration in St. Louis; the public was invited and 5000 people attended. Tesla was now more famous than even Edison.

In his lectures Tesla was able to light up a fluorescent tube without connecting any wires, transmitting the high frequency through the air. He was not the first one in the field of radio frequencies (RF); others had preceded him, notably Maxwell and Hertz.

James Clerk Maxwell was a Scottish mathematician and a generation older than Tesla. He had become enamored by Faradays images of magnetic and electric fields and was convinced that what Faraday had discovered experimentally could be captured mathematically. He worked on the idea for 20 years and wrote a book about it in **1873** [19]. In it he explained that the relationship between electricity and magnetism could be expressed in just four terse equations. More importantly, he proved that the interplay between electric and magnetic forces created "electromagnetic" waves and that light was such a wave. Furthermore, he postulated, there should be other electromagnetic waves, differing from light only in frequency.

Light has a very high frequency (around 500,000 Gigahertz). Maxwell predicted that there are electromagnetic waves at both lower and higher frequencies, traveling through space (or air) like light [9, 10]. The lower frequencies we now call radio waves (e.g. 800kHz in the AM band, 100MHz in FM); higher frequencies we have named x-rays, gamma-rays etc. The fascinating part of Maxwell's work is that no one had even noticed any radio waves.

Maxwell's book was far ahead of its time; it took 13 years to reach someone capable of understanding it sufficiently. The man was Heinrich Hertz. Six months younger than Tesla, Hertz was a professor at Karlsruhe, Germany. In 1885 he decided to investigate these alleged electromagnetic waves and he was ideally qualified for the job, having not only a thorough understanding of mathematics but also great experimental skill.

Hertz knew that, to prove Maxwell's theory, one would have to demonstrate that the alleged electromagnetic waves behaved identically to light, albeit at different frequencies. He knew he could generate different frequencies with an inductor/capacitors resonant circuit; several others had already done that. But how could he detect if a radio wave was being radiated?

He found the answer by accident. While producing oscillations with a coil, he noticed faint sparks between the terminals of a second coil which

happened to be nearby on the laboratory table and was not connected to anything [12, 21]. This gave him the idea to make a special spark-gap for the receiver, with the points very close together (he had to darken the room to see the sparks)

He chose the transmitting coil and capacitor so the frequency would be quite high (about 300MHz) and therefore the wavelength reasonably short, about 1m (3 ft). In a large lecture hall he covered one wall with a zinc sheet, reflecting the waves just like a mirror does with light. In this way he could trace the interference between the emitted and reflected waves by moving the receiver and observing how the intensity of the spark increased and decreased rhythmically along

Heinrich Hertz

the path identical to an interference pattern produced by light. He was able to focus the waves with a big saucer made from galvanized sheet metal and found that either tin foil or the body of an assistant cast a shadow [20]. Electromagnetic waves did indeed exist at other frequencies and behaved just like light.

Hertz announced his findings in **1888**, but he was not able to enjoy the fruits of his experiments. Shortly after his discovery he was diagnosed with bone cancer and was in almost constant pain. He died in 1894 at age 37.

By 1897 Tesla's living style and high-frequency experiments had consumed what money the company had received from Westinghouse and there was a problem with his $2.50 per horsepower AC royalties. The industry was in turmoil; the war between DC and AC stunted growth and royalties were not nearly as high as Tesla had expected. Edison lost control of his companies to the banker J.P. Morgan and Westinghouse was in financial difficulties. Morgan suggested patent pooling but balked at paying royalties. So Westinghouse went to see Tesla and proposed to terminate the agreement for a lump sum of $ 216,000 [25]. Tesla agreed. It was one of the worst business decisions ever made: had he insisted on collecting royalties until his patents expired, his company would have received some $ 7.5 million [21], or $ 150 million in today's dollars.

Hertz's discoveries started activity in both Europe and the U.S. In Hertz's apparatus the distance was limited by the spark-gap in the receiver, which did not respond to faint signals. An experimenter named Branly invented the "coherer", a metal tube filled with fine metal shavings. When an antenna was connected to the contacts at the ends of the tube a very small radio signal triggered a change in the way the particles conducted between themselves. The tube needed then to be shaken up for the next signal, which was a bit awkward, but the device was made to work (through the efforts of several others) by continuously tapping it with a doorbell knocker.

Tesla was well informed about these developments and was working feverishly to build more powerful radio-frequency transmitters. His resonant circuits, spark-gaps and coherers were now the most sophisticated in the world. In **1898** he demonstrated a boat that he could guide remotely without wires, turn the running lights on and off and even fire an explosive charge.

More and more Tesla now aimed his research at producing large amounts of high-frequency power. The New York laboratory became too small; to accommodate the ever-larger coils Tesla needed more room. In May of **1899** he departed for Colorado Springs, where he built an enormous high-frequency generator [29], with a coil 27m (100ft) in diameter. Connected to the coil and protruding from the building was a 46m (142ft) high metal rod. He was going to transmit a radio message to Paris.

The Colorado Springs generator was a monster to behold, producing voltages as high as 12 million volts, shooting sparks 40m (120ft) long all over the place. 100m (300ft) away arcs of a few centimeters could be drawn from any metal object and horses in the neighborhood went berserk from the tingling in their hoofs. He succeed in lighting up fluorescent bulbs with antennas 40km (25 miles) away, but he drew so much power from the local station that he caused a massive power failure and burned out a generator. He never transmitted his message to Paris.

Colorado Springs began to reveal the sad truth: Tesla had taken a wrong turn; his fabled imagery had failed him. Many believed that light (and consequently radio-waves) could not possibly travel through empty space, that there had to be an unseen substance everywhere which allowed such waves to propagate like sound in air. This magical substance was given the name "ether". Tesla not only believed in ether but was convinced it contained an infinite amount of energy, yours for the taking. In his mind it only took the right radio frequency to release this energy. Radio communication had become a secondary objective; his first goal was the transmission of power. Large amounts of power. In fact an infinite amount of power.

In January 1900 he returned to New York, having spent $ 100,000 in eight months in Colorado. He then raised $ 150,000 from J.P. Morgan to build an enormous transmitting tower on 200 acres 100 km (60 miles) east of Brooklyn at Shoreham, Long Island. The building was designed by Stanford White, the power generators ordered from Westinghouse. He had a 37m (110ft) deep well dug and over it erected a 57m (170ft) high wooden tower which carried a mushroom-shaped structure 30m (90ft) in diameter, to be clad in copper [11].

This installation, which Tesla called Wardenclyffe, was to be one of six distributed over the world. Tesla was convinced that with these six towers he could supply the entire world with power. At any place in the world, he thought, one could simply tap into the earth and draw any amount of energy, free of charge. He envisioned employing as many as 2000 people at Wardenclyffe and was sure he would soon be a millionaire.

It was an incredibly fantastic scheme, which no longer had anything to do with reality. He had raised the money by selling J.P. Morgan the idea that he would beat Marconi to a transatlantic transmission. When he had spent the money and needed more he revealed to Morgan the real purpose of Wardenclyffe. Morgan pulled the plug.

Wardenclyffe never became operational, the tower was never completed, Tesla was hounded by lawsuits for unpaid salaries and bills, some still from Colorado Springs. The machinery was being repossessed and the land sold. Tesla was bankrupt and had a complete nervous breakdown.

He was now 43 years old, at the exact halfway point in his life. For the rest of his life he produced nothing of note. He formed the Tesla Ozone Company, the Tesla Propulsion Company, the Tesla Nitrates Company, the Tesla Electro Therapeutic Company, each one a costly failure for the investors.

Each year at his birthday he would invite reporters and tell them about his new inventions and each year the claims became more fantastic. He talked about a missile he was working on which moved at 500 km per second and could destroy whole armies or fleets of warships. He claimed he could transmit energy between planets and that he had developed a death ray which could destroy 10,000 planes at a range of 400km (250 miles) [21,26].

He spoke out vehemently against the theories of Albert Einstein, insisting that energy is not contained in matter, but in the space between atoms. And he never believed in the existence of the electron.

Tesla was forced to move from hotel to hotel as bills went unpaid, each of them a step lower in stature, spending more and more time at Bryant park behind the public library feeding pigeons. He died in a run-down Times Square hotel in 1943 at age 87.

There are still many people who believe that Tesla would have been vindicated if only Morgan had let him complete Wardenclyffe. There are also many stories which ascribe which claim that he had demonstrated the power of ether, such as a 1931 Pierce Arrow car on which he had replaced the engine with an electric motor and drove it for 50 miles at 90mph, receiving power from an antenna.

But no proof for any of these claims has ever been offered, they are simply rumors. The existence of ether was disproved already in 1887 by the famous Michelson-Morley experiment and nothing since then has shown this to be false.

More fascinating is the race between Marconi and Tesla to transmit a message across the Atlantic. Had Tesla not been sidetracked by his power scheme he would have been in an excellent position to beat Marconi, having a more powerful transmitter, a larger antenna and a more sophisticated receiver.

7. Revelation

How the electron finally revealed itself
and turned out to be much smaller than anyone thought
and increasingly strange.

With the telegraph, the telephone, the light bulb and the generator, electricity became indispensable. But behind the spectacular inventions and even the solid engineering feats lay the nagging question: What is it?

There were speculations and theories aplenty, but none that added much to the age-old view of electricity as a mysterious fluid, or Benjamin Franklin's insight that "electrical matter consists of particles extremely subtile".

The discovery of this secret of nature came from an unexpected direction: gases glowing under the influence of electricity in a glass tube, the forerunner of today's fluorescent bulb.

Francis Hauksbee, an experimenter in the laboratory of the Royal Society, had observed the phenomenon as early as **1705**. His job was to build instruments for the members and to assist in the experiments. Hauksbee evacuated a glass jar and, opening the valve, let mercury droplets stream into the jar. Inside the jar the mercury droplets displayed a brilliant shower of light.

But Hauksbee then found that he really didn't need mercury droplets. He evacuated a glass globe, rotated it with a crank and when he held his hand to the glass, the inside of the globe glowed. The light was bright enough to read large print and faintly illuminate the nearby wall.

The vacuum pump Hauksbee used was very poor. Up to that time a vacuum pump was very much like a water pump. The seals for the piston were made of leather; inevitably air leaked around the seal, about 4% of the air was still left inside the globe. For something interesting to happen a much better vacuum was needed, which would not happen for 150 years.

In **1855** Johann Heinrich Wilhelm Geissler came up with a new idea for a vacuum pump. Geissler was a glass blower and mechanic who worked for the University of Bonn. He filled a glass vessel with mercury and connected it through a flexible tube and a stopcock to the chamber to be evacuated. Holding the vessel high, the tube was filled up to the stopcock with mercury. Then he lowered the vessel and opened the stopcock. Now the weight of the mercury pulled some of the air out of the chamber. The stopcock was then closed and the procedure repeated many times, lifting the vessel, filling the tube, lowering the vessel and opening the stopcock.

Geissler's way of pulling a vacuum may have been excruciatingly slow, but the mercury formed a much better seal and he managed to achieve a vacuum of 0.01% of atmospheric pressure [3]. He evacuated long glass tubes with two metal electrodes sealed through the walls at opposite ends and found that, the lower the air pressure inside the tube the more interesting the glow became when the electrodes were connected to a high-voltage battery or electrostatic generator. Geissler was, in fact, improving on an experiment that Michael Faraday had performed in 1838. Faraday had observed a purple glow extending from the positive electrode (which he named *anode*) to near the negative electrode (the *cathode*) [1].

Julius Pluecker was the professor at Bonn for whom Geissler did much of his glass blowing. Pluecker became fascinated by Geissler's tubes and decided to investigate the phenomenon in more detail. When he modified the design so that the glow became a narrow capillary in the middle, he discovered that a magnet could deflect this light stream. Pluecker published his findings in **1858** and thus stimulated the interests of other investigators. The question now became: what did this stream of light consist of?

Progress in identifying the phenomenon went hand in hand with improvements in vacuum technology. The next major step was taken by Hermann Sprengel, who was born in Germany but worked in England. Sprengel invented the mercury-drop pump in **1865**, which eliminated the need to raise and lower a vessel. With his pump the vacuum was now down to 0.001% of atmospheric pressure.

Next came Eugen Goldstein who, after receiving a Ph.D. from Berlin, spent most of his life at the Potsdam observatory. It was Goldstein who first named the phenomenon cathode rays, recognizing it was the cathode that produced the mysterious emission. His most important discovery was that these cathode rays came at right angles out of the surface of the cathode. They were not like light from a piece of glowing coal, for example, which shines in all directions. Cathode rays were projected only perpendicular to the cathode surface.

Now it was no longer an investigation of glows, but one of a ray (or beam), which produced fluorescence where it hit the wall of the glass tube. In **1871** Goldstein made a slightly concave cathode and with it was able to focus the cathode beam into a bright spot on the glass envelope. The fluorescent spot on the glass could be moved with a magnet placed anywhere along the beam. The spot became brighter and better focused as more air was pumped out of the tube.

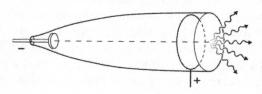

Cathode-ray tube.

So far the glamorous part of the research on cathode rays had been done in Germany. But in 1878 an English investigator, William Crookes, appeared on the scene and shifted the focus away from Germany.

Crookes worked mostly outside established research institutions. He was born in 1832, the son of a prosperous tailor with 15 other children. After some irregular schooling he studied at the Royal College of Chemistry in London and made contact with Faraday and Wheatstone. In many ways Crookes was similar to Faraday: both men were brilliant experimenters and lecturers, well organized and orderly in their procedures and largely ignorant of mathematics.

After a stint at the Radcliffe Observatory at Oxford and a teaching job at the College of Science at Chester, Crookes settled in London. His father's wealth made him financially independent; he worked occasionally as a chemical consultant, edited photographic and scientific journals and experimented in his own laboratory.

In **1878** Crookes started his own research on cathode rays. He quickly realized that drawing a nearly perfect vacuum was the key to the mystery and he managed to improve the Sprengel pump to the point where he could achieve a vacuum of one millionth of an atmosphere. With a sharply focused beam he was able to heat metal foil to a red glow. He found that the beam, directed toward a metal object, charged it with a negative potential. From these observations Crookes concluded that the beam must consist of negatively charged particles, which he called a "torrent of molecules" [12]. In Crookes' view these molecules broke loose from the cathode and, through friction, produced the glow in the gas and the glass envelope.

Twelve years later, in **1891**, Heinrich Hertz came to a different conclusion. He, too, had directed cathode rays toward a thin gold foil in an evacuated tube, but he found that a portion of the beam penetrated the foil and emerged on the other side in a diffused form. Particles (meaning atoms or molecules), Hertz argued, could not penetrate solid metal; cathode rays, therefore, must be a wave similar to light [2].

One of Hertz's assistants, Philip Lenard, produced additional evidence for Hertz's argument. In 1892 Lenard managed to mount a thin aluminum foil into an opening of the glass envelope. Directing the cathode beam against this window he discovered that a portion of the beam came through into the open air. For up to two centimeters (0.8 inches) outside the aluminum foil a photographic plate became exposed, even though no light was present [12].

The aluminum foil was certainly very thin, but it was thick enough that not even the smallest known atom or molecule could penetrate it. This, to Lenard and Hertz, was conclusive proof that the cathode did not emit a ray of particles, but a wave akin to light [1].

For the next few years a battle of words raged over the nature of cathode rays. Almost all German scientists sided with Hertz and Lenard, almost all English ones with Crookes. The evidence was insufficient and controversial. Light waves could not be deflected with a magnetic field, but cathode rays were. On the other hand, no particle had ever been observed penetrating a metal foil.

In **1895** the investigation of cathode rays took a surprising and spectacular turn as a researcher new to the field, Wilhelm Roentgen, repeated some of the earlier experiments. What Roentgen discovered didn't add anything to the knowledge of the nature of the beam, but it lead to an entirely new phenomenon.

Wilhelm Conrad Roentgen was born in 1845 in Lennep, Germany. His father was a cloth manufacturer and merchant. When he was three years old the family moved to Appeldoorn in Holland. Roentgen attended a private boarding school, where he was judged not particularly studious. At age 17 he went to the Utrecht Technical School, but was expelled when he was among a group of students standing around and laughing at a caricature of an unpopular teacher. Roentgen was the last one to disappear when the teacher entered the classroom, was caught, and refused to identify the classmate who had drawn it [10].

Because of this expulsion Roentgen had trouble being accepted as a regular student. In desperation he applied to the newly formed Polytechnic in Zurich, Switzerland, hearing that this school didn't care so much about the student's background as long as he passed an entrance examination. Roentgen passed, attended for four years and graduated as a mechanical engineer. But then he decided on an academic career in physics. He stayed on, got his Ph.D. and worked as assistant to professor August Kundt.

Wilhelm Roentgen

In 1870 Kundt moved to the University of Wuerzburg and Roentgen moved with him, still an assistant. Two years later Kundt moved again, this time to Strassburg, Alsace (then Germany, now Strasbourg, France); Roentgen moved with him, still an assistant because of his irregular educational background and lack of classic languages (which were then considered an essential part of education). In 1879 he finally became a full professor at Giessen, where he stayed for nine years and then moved to the University of Wuerzburg in Bavaria as professor and director of the physics institute. In 1894 he also became Rector of the university [10].

Roentgen was a shy man who didn't talk very much. His lectures were exceedingly dull; he spoke in a soft, monotonous voice and was hard to hear in the back rows. Yet he was a stern disciplinarian and his examinations were dreaded.

Roentgen's love and forte was experimental physics. In the laboratory he was a perfectionist, driving himself hard to squeeze out the last bit of precision, always working alone. Up to 1894 he had published 48 scientific papers, solid, diligent, painstaking work, but all of it un-dramatic. It was his 49th paper that would change the world.

On November 8. **1895** he noticed something unusual. With his cathode ray tube draped in black cardboard and thus allowing no light to escape, he saw something light up nearby. It was a screen, painted with some exotic crystals, which happened to be on the bench about a meter (3 ft) away. When Roentgen turned off the voltage to the cathode rays tube, the illumination disappeared [12].

With his customary diligence Roentgen investigated the effect. Invisible rays, he found, emanated from the luminescent spot where the cathode rays hit the glass wall of the tube. They traveled about two meters

(6 ft) in air and could not be reflected, refracted or deviated by a magnet. Most astoundingly they penetrated materials, which were opaque to light.

Roentgen grew nervous. He wasn't sure whether his observation was real or merely an illusion. He was in very bad humor, talked even less than usual, ate little, didn't answer his wife when she asked him what was wrong, disappeared into his lab immediately after dinner and even slept there on a cot.

Roentgen was colorblind and didn't trust his eyes to judge the phenomenon. He sought confirmation with photographic plates and found it in a spectacular way. With the plate completely wrapped in black paper so that no light could penetrate, he was able to photograph metal weights inside a closed wooden box and the loaded chamber of a shotgun. The strange rays were stopped by dense, thick objects, but light or thin materials were transparent to them. On one of the photographs he saw the bones of his own hand; the rays had penetrated the flesh but not the bones.

On December 22, 1895 Roentgen showed his discovery to his wife Bertha, taking a picture of her hand (exposing it for 15 minutes). Then he wrote an account of his investigation and sent it to the editors of the Physical and Medical Society of Wuerzburg. He named the phenomenon X-rays (for X, the unknown in mathematics) and speculated that it was some sort of ultraviolet radiation. He was certain that these rays were not cathode rays but created by them in the glass wall.

Roentgen's paper was hurriedly printed and he sent copies, together with some photographs, to his professional acquaintances. The photograph of Bertha's hand shook them, the eerie picture of a living skeleton. The Wiener Presse learned of the discovery and, on January 15, 1896, printed the story. In the following days Roentgen's discovery appeared in newspapers all over the world.

Roentgen's quiet existence as a professor was at an end; fame was to haunt him for the rest of his life. On January 13 he was summoned to demonstrate X-rays to the Kaiser in Berlin. Within three months five editions of his paper had to be printed and a whole army of physicists and experimenters started taking X-ray pictures. During the year 1896 alone some 1,100 books, pamphlets or papers were published on the new subject and, although Roentgen made no patent claims, soon applications started pouring into patent offices in Germany, France, England and America [12].

The reaction of the public ranged from fear to wild misunderstanding. Almost immediately Roentgen received letters criticizing him for spreading the discovery of these "death rays" which could destroy all mankind [10]. Only six weeks after the announcement, Assemblyman Reed of Somerset County, New Jersey, introduced a bill into the house prohibiting the use of X-rays for opera glasses in theaters. A quick-reacting, profit-minded firm in London advertised the sale of X-ray-proof underclothing [5]. On the religious scene Roentgen's X-rays proved the existence of a spiritual body in man, according to the Herald and Presbyter. The New York magazine Science printed a story describing how anatomical diagrams had been projected directly into the brains of advanced medical students with X-rays, an approach supposedly much better that ordinary teaching methods, and a Cedar Rapids, Iowa, newspaper reported that George Johnson, a young farmer, had been able to transform cheap metal into gold with X-rays [10].

Roentgen's discovery had done nothing to explain the nature of cathode rays, in fact it had added to the confusion. Soon it was found that cathode rays striking a metal surface produced even stronger X-rays than glass. Now the question was whether Lenard's aluminum window had proven anything at all or whether Lenard had merely produced X-rays and never recognized them as such.

The riddle was at least partially solved by an Englishman, J. J. Thomson, a professor at Cambridge and head of its Cavendish Laboratory. He was a slight, unassuming man, a bit forgetful and rather clumsy with his hands (his laboratory assistant begged him not to handle the equipment himself).

He was dissatisfied with the bickering over the nature of cathode rays and sought a more definite experimental proof. Being in the English camp, however, he was quite convinced that such a proof would show that cathode rays were small particles.

J.J. Thomson

If Thomson's speculation were right, such particles would have a mass (or weight), carry a small negative electric charge and move with a certain velocity. The mathematical relationship between mass, charge and velocity of such a hypothetical particle was quite easy to figure out. Thus, if he could measure two of these unknowns, the third one could be calculated.

His first attempt was to measure the velocity, using rotating mirrors. Although the result was clearly inaccurate, the velocity of the beam always turned out to be much slower than the speed of light, strengthening the argument that the beam was not a wave [1].

Thomson was unable to measure either the mass or the charge of these supposed particles directly, but he found two independent ways to determine at least the ratio of these two factors. In **1897** he measured the amount of heat produced by the beam on a target and thus got an idea of the kinetic energy. Next he measured the electric charge the beam transported to a metal object. And finally he measured the amount of deflection produced in the beam by an electrostatic voltage and how strong a magnetic field was required to bend the beam back into a straight line [1].

The first two or the last two were sufficient to determine the ratio; when the two sets of data gave the same results, the evidence for a particle far outweighed that for a wave. But it was a particle never seen before, more than 1000 times smaller than the smallest atom, hydrogen. And the same small particle showed up for every substance Thomson investigated, leading to the conclusion that this was a part of every atom and the atom, contrary to its name, was not indivisible. Thomson, at long last, had found the electron.

The next step came from two of Thomson's students at the Cavendish Laboratory at Cambridge. C.T.R. Wilson invented a cloud chamber, a box filled with moist air, so moist, in fact, that it was close to raining inside. With a piston connected to the chamber Wilson could suddenly decrease the air-pressure and make the moisture condense in droplets (i.e. it began to rain inside the box). In doing this, Wilson noticed that the first droplets always started at a dust particle or an electrically charged molecule (an ion). It was this second feature, which was of great interest.

John Sealy Townsend drew this electrically charged mist with a suction tube into an electrically insulated box, let it dry and then measured the charge. By estimating the number of droplets by the volume of water and the rate of fall of a drop (which gave him the approximate size) he was able to crudely estimate how much charge each electron carried [1].

J.J. Thomson refined this method by ionizing (charging) the air with X-rays. In this way he could observe large numbers of charged droplets in the cloud chamber. These measurements got Wilson, Townsend and Thomson to within about 30 percent of the correct charge.

The scene now moves to America, to the University of Chicago. Robert Millikan was born in 1868 in Illinois, the second son of a Congregational minister and grew up in a small town in Iowa. In high school he did well in mathematics, but wasn't particularly interested in science. He went to Oberlin College and at the end of his sophomore year his Greek instructor asked him to teach an introductory course in physics, not because he was good in physics, but because he had done well in Greek.

After graduating, Millikan's first inclination was to become a gymnast (he was outstanding at the parallel bars), but he suddenly realized that he

had had the greatest fun teaching the physics course and he stayed on to study for a master's degree [8]. In 1893 he went to Columbia University on a fellowship (and found he was the only graduate student in physics). After he earned his Ph.D. he toured Europe (traveling some 6000 km (4000 miles) on a bicycle) and studied at Jena, Berlin and Göttingen. Then he received an appointment as assistant professor at the University of Chicago, his home for the next 27 years [8].

Millikan was a born teacher. He got absorbed in his job and, finding a dearth of good physics

Robert Millikan

textbooks, he wrote five of them, at four different levels. For the first ten years in Chicago he had little time to do research and, for that reason, it took him 11 years to become an associate professor. Realizing that he might never make it to full professor, he spent more time on research and chose a really tough problem: the measurement of the electron charge.

Millikan began in **1909** by setting up his own Wilson cloud chamber. He mounted two large parallel plates into the chamber, one at the bottom, the other at the top, with the idea of pulling the charged droplets up by electrostatic attraction, counteracting gravity. But when he turned on the voltage - 10,000 volts of it - the cloud disappeared in a puff. The experiment looked like a total failure until he noticed, through a short-focus telescope, that a few droplets had remained in the middle. When he increased the voltage, they moved up, pulled

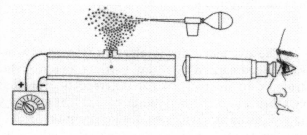

Millikan's oil-drop experiment, with which he determined the properties of a single electron.

by the positive upper plate: when he decreased the voltage they began to slowly fall, pulled by gravity.

One of the sources for inaccuracy in these measurements was the evaporation of water; during the time of the measurement the droplets became smaller and smaller. It occurred to Millikan that clockmakers had, for hundreds of years, been improving fine lubricating oil so that it would not evaporate for a long time. He now changed his experiment, drilling a small hole in the upper plate and spraying oil with an atomizer above it. Every so often an oil droplet would fall through the hole. He could then ionize it with X-rays and, watching it through the telescope, keep it suspended for hours by adjusting the voltage. By measuring the time it took for the droplet to fall a precise distance (two crosshairs in the telescope) and knowing the viscosity of air, he was able to determine the size and weight of the droplet and the amount of charge it held.

Exhilarated by the find Millikan spent long hours at the laboratory, measuring many different droplets and correcting the law governing the fall of bodies through gas of the oil-drop. At one point he had to call his wife and tell her that he would have to miss a dinner party arranged for that evening because he had watched an ion for an hour and a half and had to finish the job. One of the guests was horrified that associate professors were so poorly paid that Millikan had to "wash and iron" for an hour and a half and could not attend the dinner party because he had to finish the job [8].

By measuring hundreds of droplets Millikan made a fundamental discovery: the amount of charge on each droplet always appeared in discrete jumps, depending whether the droplet had collected one, two, three or more electrons. There was never any value in between. He now realized that he was staring at a single electron for the first time, or at least the manifestation of it.

The existence of the electron was now widely accepted. Electricity was indeed carried by a tiny particle, but how did this particle fit into the structure of matter? To get a perspective on that question we must step back briefly to the beginning of the 19th century, when Dalton made the first significant advance in the recognition of the structure of matter since Grecian times. John Dalton, a self-taught scientist, had an interest in meteorology, specifically the theory of rain. It was the common belief that water was completely dissolved in the air, i.e. that the air was a vast chemical solvent. Dalton, through painstaking weather observations, became convinced that this was not so, that water always existed separately, diffused in the air. This led him to the theory that, in a mixture of gases, every gas acts as an independent entity. He noticed that gases always combine in a certain ratio; one ounce of hydrogen, for example, always mixes with eight ounces of oxygen to form water.

By 1803 Dalton was quite certain that gases, liquids and solids consisted of ultimate particles and that any chemical combination depended on the weight and number of the respective particles. In his view these chemicals never mixed fully, but only combined in certain fixed ratios.

Dalton read a paper before the members of the Literary and Philosophical Society of Manchester (with a grand total of some nine members in attendance) and then was invited to give two lectures at the Royal Institution in London. At first his theory was not widely accepted; Davy and his peers were not convinced. But Dalton kept working on the concept, testing the weights and mixing ratios of more and more gases, giving lectures and publishing a two-volume work. Gradually he won over more and more of the chemists to his view by the simple fact that his theory worked: it could predict how gases (and later liquids and solids) combined.

Dalton had chosen the word atom for his ultimate chemical particle, reviving an ancient Greek speculation of the indivisible basic structure of matter. At first he used the term atom loosely, generally for an element, but sometimes even for compounds. Later he began to speak in terms of indivisible elements and finally became dogmatic about it: "I should apprehend there is a considerable number of what may be called elementary principles, which can never be metamorphosed, one into another, by any power we can control". Of course, Dalton was right as far as the knowledge of chemistry in the early 19th century went. Despite the alchemists, no element had ever been observed to change into another, and no such observation would be made for almost 100 years.

The next major step in the knowledge of the atom came in 1869. Dimitry Ivanovich Mendeleyev (or Mendeleev) was a professor of chemistry at the University of St. Petersburg. While working on a textbook it suddenly occurred to him that the listing of elements simply by atomic weight didn't tell the whole story. Similar chemical properties seemed to repeat themselves periodically as one went down the list. Mendeleyev rearranged the elements in a "periodic table", listing the elements with similar properties underneath.

But why would seemingly unrelated elements have similar properties? Mendelyev's periodic table suggested that there was more to the atom that the simple, inert, indivisible particle Dalton had pronounced.

In 1896 Roentgen's pamphlet on the new kind of rays was particularity intriguing to Henri Antoine Becquerel. Becquerel was a professor at the Museum of Natural History in Paris and had done considerable work on fluorescent crystals. What intrigued him was the fluorescence of the glass, the spot that actually produces the X-rays. He wondered if fluorescent crystals might not also emit such rays. Having learned of Roentgen's discovery on January 20. 1896, he reported to the Academy of Sciences already on February 24 that he had indeed been able to get patterns of exposure on photographic plates from some crystals made to fluoresce in sunlight.

The following week was overcast in Paris and Becquerel, having already prepared more experiments, stored the crystals in a drawer, right on top of the photographic plates. The sun didn't come out all week, but for some unknown reason Becquerel decided to develop the plates anyway. He found the plates exposed; it was thus not the fluorescence, which caused the exposure of the plates, but something inside the crystals.

Now he tried various compounds and discovered that the exposure was strongest with pure uranium. It was the uranium, which continuously emitted some kind of radiation [1].

Marie and Pierre Curie, also working in Paris, followed up Becquerel's discovery. Were there other elements, besides uranium, which radiated? They found that thorium was also "radioactive", as Marie Curie termed it and, after an arduous separation process, found the radioactive element radium.

These three elements emitted a faint amount of energy. Something was coming from inside these atoms, again a strong suggestion that the atom - or at least some atoms - were not nearly as simple as Dalton had wanted them to be.

Now came upon the scene a remarkable man who provided not just the next answer, but a whole series of them. Ernest Rutherford was born in New Zealand and, with the help of scholarships, received a B.A. and M.A. His first interest was radio transmission; he invented a sensitive detector and, when he went to Cambridge in 1895, again on a scholarship, he continued working on it. But in 1896 J.J. Thomson asked him to join in some research on the influence of X-rays on gases, and Rutherford soon found that it was research, not invention, which truly interested him [11].

Ernest Rutherford
(Observe the sign "Talk Softly Please"; Rutherford was well known for his booming voice).

The research with X-rays bore fruit almost immediately. When Rutherford heard of Becquerel's discovery he compared ionized particles produced by radioactivity to those produced by X-rays and found them to be the same. But he also found that, in the case of radioactivity, there were really two rays, a strong one, which caused a positive charge, and a weaker but more penetrating one, which produced a negative charge. Rutherford named them simply alpha and beta rays.

Unfortunately, the possibilities for advancement and earning more money were dismal at Cambridge. Rutherford accepted an offer from McGill University in Montreal and there he made a stunning breakthrough. Together with a young chemist, Frederick Soddy, he set out to investigate the nature of these mysterious radiations. Isolating beta rays, they found them to be particles with the exact properties of electrons, in fact there could be no doubt that beta rays were simply electrons shooting out of radioactive materials at high speed.

The measurement of the alpha rays turned out even more stunning. These, he found, were particles with the mass of a helium atom. Ever since J.J. Thomson's proof of the existence of the electron it had been suspected that the electron was somehow part of the atom, with the number of electrons per atom increasing with the atomic weight of the element. But here was a helium atom, which apparently had been stripped of its (two) electrons, leaving it with a positive charge.

But why would a material like uranium constantly send out electrons and incomplete helium atoms? Where did this energy and mass come from? Rutherford postulated a daring theory: uranium was slowly transforming itself into lead (or, more accurately, an isotope of lead), an element of lower atomic weight and, in doing so, was giving up energy and mass. What Dalton said couldn't be done was in fact happening.

In **1907** Rutherford moved back to England, to the University of Manchester, now a well-paid professor. A year later he received the Nobel Prize for chemistry. With Hans Geiger (who later became well known for his Geiger counter) and Ernest Marsden he continued to probe into the atom. Rutherford recognized that alpha particles might be useful miniature projectiles. Geiger and Marsden set up an experiment, directing an alpha ray against a thin foil of gold. Not much should have happened; J.J. Thomson had pictured the atom as a uniform sphere in which the electrons were imbedded, like raisins in a pudding. The dense gold should have slowed down most of these particles, perhaps deflected some by a few degrees. But a few of the alpha particles bounced straight back, as if they had hit a massive object [1]. It was like firing a rifle at a sheet of plywood and finding that some of the bullets came straight back -- at high speed.

Rutherford knew almost immediately what this meant. The gold foil did not consist of uniform blobs of atoms, the mass of the atom was concentrated in a very small, dense, and positively charged nucleus. Electrons were located around this nucleus, giving the atom its apparent size. Between the nucleus and the electrons there was nothing but empty space. Once in a while an alpha particle hit the much more massive nucleus of a gold atom and bounced right back. All other alpha particles moved straight through the empty space or were deflected a little if they came too close to a gold nucleus.

Next question: how did the atom hold together? At first it was thought that the electrons were distributed evenly around the nucleus in stationary places. Attracted by the positive charge of the nucleus they would move inward until their mutual repulsion kept them at some distance. But Rutherford came up with a different answer: like planets in a solar system, the electrons are orbiting the nucleus. It was a brilliant idea but it really produced more problems than solutions. From everything that was understood about moving charged bodies, such orbiting electrons should have radiated all the time, thus giving off energy. As their energy decreased they should have slowed down and gradually spiraled into the nucleus [1]. Rutherford, ever self-confident, wasn't bothered too much by this inconsistency; his intuition told him that he was on the right track.

In **1912** Rutherford was visited by a young Dane, Niels Bohr, who had obtained his Ph.D. in Copenhagen and had come to England to work with J.J. Thomson. But Thomson somehow didn't show much interest in

Bohr's work and Bohr latched on to Rutherford. They were as different as any two men could be. Rutherford, a large, exuberant man with a booming voice, was given to uncontrollable excitement and worked with immense gusto [14]. Bohr, punctilious and excessively sensitive, spoke in a whispering voice, sometimes pausing for minutes trying to find the right word. Yet the two became immediate friends and remained in contact all their lives [13].

It was clear to both Rutherford and Bohr that a model of the atom would have to explain the emission of light. Already back in the 19th century the field of spectroscopy had grown to considerable sophistication. Each of the elements emits sharply defined characteristic wavelengths (or frequencies) of light when heated. Since light had already been recognized as electromagnetic radiation, there was little doubt that these emissions must be caused somehow by the electron, i.e. by a rapid movement of its charge.

Niels Bohr

Bohr studied Rutherford's model and quickly proposed a modification. The electrons, he said, can orbit only at distinct distances from the nucleus, in certain stable, spherical paths or shells. They tend to go to the lowest state of energy, the innermost shell. When energy is supplied to the atom, say in the form of heat, the electrons jump to one of the outer shells. When they jump back again they emit the characteristic frequency.

Bohr was able to calculate the stable orbits for hydrogen. When he figured out the frequencies created by transitions between all the combinations of orbits, it was an exact fit to the observed spectrum of hydrogen.

But almost immediately after Bohr explained the inner workings of the atom it became clear that this was not the end of the story. Bohr's electron orbits explained hydrogen well, but it failed quite miserably for heavier elements.

In 1923 Louis de Broglie came up with a refined theory. In his mathematical calculations he assumed the electron to be a *wave* rather than a particle and immediately got a much better fit with the observed behavior.

We now know that the electron only *appears* to be a hard particle, but is fundamentally a wave. For mere humans this is very difficult to comprehend and impossible to imagine. The concept can only be grasped through abstract mathematics.

The main difficulty in particle physics is our language. We speak in metaphors - we explain things unfamiliar to others in terms of common experiences. But when it comes to a particle so small that none of us are able to experience it, language breaks down. In the realm of the very small, physical laws as we experience them no longer apply. It is a strange and perhaps forever unimaginable world.

And what about magnetism, the early and constant companion of electricity? It turned out to be caused by the electron, as Ampere had suspected. The motion of an electron creates a circulating current, which in turn created a magnetic field. In most gases, liquids and solids the orientation of the electrons is random so that these small magnets cancel each other out. In ferromagnetic materials (such as iron) an external magnetic field forces the electrons to align themselves. In permanent magnets the atoms are so close together that they influence each other and the electrons remain in common alignment, producing a magnetic field in unison.

William Crookes crowned his illustrious career with a knighthood and by being elected president of the British Association for the Advancement of Science, the Chemical Society, the Institution of Electrical Engineers and the Royal Society. Although illustrious, Crookes' professional path left a few doubts. Just before his work with the cathode rays he became involved in spiritualism. Having supposedly made a scientific investigation he defended an 18-year-old medium, Florence Cook, as genuine. After his death in 1919 evidence came to the surface that Crookes and pretty little Florence had had an affair and the "scientific investigation" was little more than a cover-up.

Wilhelm Roentgen was annoyed by the public's misunderstanding of his discovery, but he didn't think there was any point in science-education for the average citizen. He became bitter at the intrusion into his privacy and the rival claims. After two more papers on his initial research on X-rays he wrote nothing more on it and published only six more papers during his life.

In 1900 he moved to the University of Munich (where he was loaded down with administrative work) and in 1901 he received the Nobel Prize for physics (the first one offered). He retired in 1920 at age 75 and died in 1923, having stipulated in his will that all records of his work between

1895 and 1900 and all correspondence concerning the discovery be burned unopened [10].

For Roentgen there was little joy in living the last few years of his life. He experienced the defeat of Germany in World War I, witnessed the communist upheavals in Munich, and was gripped by the economic collapse. He had bequeathed his Nobel Prize money to the University of Wuerzburg, only to watch it become worthless in the hyperinflation. His hometown of Lennep issued emergency money with Roentgen's picture on it; the face value of the banknote was one billion marks [12].

Philip Lenard received the Nobel Prize in 1905, but his career took a bad end. He had always felt neglected and pushed aside - even by Hertz - and this hypersensitivity became pathological. He felt that he, not Roentgen, was the true discoverer of X-rays and embarked on a campaign to belittle the work of Roentgen and make him appear a bungler who had made a discovery by accident [6]. In his publications he either ignored Roentgen or dismissed him in a footnote. Later Lenard attacked J.J. Thomson in a similar way, claiming priority in the discovery of the electron.

In 1919 Lenard called for rearmament of Germany in his lectures. In 1920 he delivered a vicious attack on Einstein, dismissing relativity as "dogmatic Jewish physics". By 1924 Lenard declared Hitler a "true philosopher with a clear mind". In 1935 he pronounced that he had detected a split personality in Hertz, whose theoretical works were the product of his half-Jewish inheritance. In his four-volume 1936/37 work entitled "German Physics" there is no mention of either Einstein or Roentgen and he claims that science is racially determined. Needless to say, Lenard became the darling scientist of the Nazi regime. He died in 1947, having witnessed its defeat.

J.J. Thomson's received the Nobel Prize in 1906. Under his leadership the Cavendish Laboratory became one of the world's leading research institution in physics. He was an outstanding teacher (seven Nobel prize winners among his students) and he supervised the construction of new buildings for the laboratory, financed entirely by student fees [15, 16]. He was knighted, became Master of Trinity College and remained in this position until his death in 1940 at age 83.

Ernest Rutherford received the Nobel Prize in 1908. During World War I he worked on submarine detection, but then went on with research in his main field. In 1919 he accomplished the first artificial disintegration of an atom, using alpha rays to convert a nitrogen atom into one atom of hydrogen and one atom of oxygen. The same year he succeeded J.J. Thomson at the Cavendish Laboratory. In 1932 he became Lord Rutherford of Nelson. He died in 1937, at age 66.

Niels Bohr returned to Denmark and became director of the Institute for Theoretical Physics in Copenhagen, a post he held until the end of his life. He received the Nobel Prize in 1922. In 1943 he escaped to Sweden from the Nazi-held Denmark and from there went to England and Los Alamos. Even before the atomic bomb was completed he wrote letters to Churchill and Roosevelt, urging international control of nuclear weapons. He died in 1962, age 77.

Robert Millikan moved to California in 1921 as chairman of the executive council of the then small California Institute of Technology and received the Nobel Prize in 1923. He continued teaching and directing research, but he now showed an uncanny ability as a fundraiser. It is primarily due to his efforts that Caltech became a university of such high standing. Millikan officially retired in 1946 at age 78 but continued working until his death seven years later.

8. Armstrong

Marconi's doubtful transatlantic transmission,
the man who invented the vacuum tube but didn't understand it,
the genius who made radio work
and his deadly adversary, a ruthless self-promoter.

Everybody knows that Marconi invented radio. Yet history tells us something quite different: most of the credit went to the wrong people.

Guglielmo Marconi's father was a fairly wealthy landowner in Bologna, Italy; his mother was the daughter of a Scot who had built a distillery in Ireland. Their marriage was not a happy one. Father Marconi was a dour, parsimonious provincial; his wife was an elegant lady who wore daring hats and spent the winters in fashionable resorts, taking her young son along and relying on tutors to educate him. When Guglielmo finally attended a regular school, he was found backward for his age and a bit "stuck-up". To protect himself from an often tyrannical father he became increasingly withdrawn and self-reliant [29].

Marconi flunked the entrance exam to the University of Bologna, but mother charmed a Professor Righi into taking the boy under his wing. Righi's field was physics and by then the young man had developed a keen interest in things electrical [22].

In the summer of 1894 Righi introduced Marconi to the work of Hertz. Marconi couldn't believe that no-one had developed these findings into a commercial product. He read everything he could find on the subject, including work done after Hertz. He had no interest in fundamental theory, his one and only goal now was to build a transmitter and receiver which would transmit Morse code [1, 22].

In the spring of 1895 Marconi started experimenting, secluding himself in the attic of the country villa. He duplicated Hertz's transmitter and a "coherer" designed by the Frenchman Branly and soon found that a mixture of 95 percent nickel and 5 percent silver filings in the coherer provided the most sensitive reception. As an indicating device he used an electric bell, which was turned on every time the coherer fell into low resistance by the arriving signal. He even had the ingenious idea of shaking the coherer with the hammer of the bell so that it would be automatically restored and ready for the next signal [1].

By the spring of 1895 Marconi started to experiment outdoors. He had noticed that the transmission distance increased every time he made Hertz's antenna wires longer and added larger metal balls or plates at the ends (thus decreasing the frequency). Struggling with the large and awkward contraption, which had to be hauled up tall poles, he found that using only one end would work quite well if the second wire was stuck into the ground. Tesla had discovered the same thing four years earlier. But Tesla was after power, Marconi's aim was distance: he was able to transmit over 2.4 km (1.5 miles) [1].

Marconi now felt ready to make some money and offered his invention to the Italian Ministry of Post and Telegraph. There was no interest. With the encouragement of his mother he decided that Great Britain would be more receptive to progress and the two of them left for London. It appears that the father was more than glad to get rid of the mess of wires and coils, which never made any commercial sense to him; he had been destroying Guglielmo's equipment whenever he found it lying around.

In London, Mrs. Marconi sought the help of a well-connected nephew, who got an introduction to William Preece, chief engineer of the Government Telegraph service. In June 1896 Marconi applied for a British patent and in July he made his first demonstration over a few rooftops in London. In September the Army and Navy attended and in December William Preece gave a public lecture on Marconi's system, during which Marconi walked around the audience with a receiver, which rang whenever Preece pushed a telegraph key at the podium. For dramatic effect both transmitter and receiver were sealed in mysterious black boxes [6].

The press carried the story, crediting Marconi with far more than he had actually invented [6] and the 22-year old became an instant celebrity; he was now possessed by a Messianic belief in a great concept. By March **1897** he was able to bridge 7 km (4 miles) with wire-antennas held aloft by balloons and by summer he managed to transmit over 14 km (9 miles) across the Bristol Channel. In July he raised £ 40,000 of venture capital and founded the Wireless Telegraph and Signal Company Ltd. (later renamed the Marconi Wireless Telegraph Company).

Marconi was in business, but profits turned out to be elusive. What attracted investors was the expectation of an imminent boom in wireless, similar to - or perhaps even bigger than - the boom created by the telephone. But Wireless was competing against both the telegraph and the telephone; its only chance for profit was to reach where wires couldn't: over the ocean, spanning large distances.

Marconi steadily increased transmitter power. In 1898 he was able to report the results of a regatta in the Irish Sea from a boat 32 km (20 miles) from shore. By 1900 he bridged a distance of nearly 100 km (60 miles) between two British warships [6].

Now Marconi made plans for the one demonstration that would convince everyone of the capability of wireless: a transmission across the Atlantic. The venture capital had allowed him to hire 17 professional engineers and to employ the best available consultants: Lord Kelvin and Professor John Ambrose Fleming. Fleming designed the transmitters for the two transatlantic stations, each powered by a 25 kilowatt generator coupled to a 32 horsepower steam engine. Two sites were selected: Poldhu on the west coast of Cornwall and Cape Cod, Massachusetts. Construction started in October of 1900. Each antenna was to consist of 100 huge masts, arranged in a circle. But the construction was faulty. By August 1901 the antenna on Cape Cod was seriously distorted by high winds. In September all 100 masts at Poldhu came crashing down in a gale; a few days later the ones at Cape Cod collapsed too [23].

Marconi went into an emergency mode; he had already announced that he would transmit across the Atlantic before year's end and received enormous publicity. He had a temporary antenna erected at Poldhu, using two 50-meter (150-ft) masts; the transmission would only be one way, from Poldhu to a new site in Newfoundland, which was a bit closer. The receiving antenna in Newfoundland would simply be a wire, pulled 130 m (400 ft) into the air by a kite [23].

He arrived in Newfoundland on December 6, **1901**. On December 9. he sent a telegram to Poldhu with instructions to send the letter S - three dots - every day between 11 am and 3 pm, Newfoundland time, starting on December 11.

On December 11. there was a gale and the kite broke loose. On December 12. the high winds persisted, the kite soared and dipped erratically and nothing was heard. He tried a different receiver and then, at 12:30 pm, both he and his assistant heard the three dots briefly in the earphones. The following day they were again able to hear them for a brief period [23].

Guglielmo Marconi

There was a problem, though. Marconi had not been able to make his own receiving equipment work, a receiver which drove an inker, giving proof of the reception. The only receiver that worked had been given to him by the Italian Navy and had earphones. Nevertheless, having received no more signals after December 13., he decided to announce the accomplishment to the press.

Most electrical engineers and scientists were skeptical. It was almost universally believed that electromagnetic waves, traveling in a straight line, could not be received beyond the horizon. Between England and Newfoundland the earth forms a hump more than 200 km (125 miles) high and it was thought very unlikely that waves would travel either through this huge mountain or bend over it.

As it turns out Marconi was incredibly lucky. Triggered by Marconi's claims, two men, A.E. Kennelly in the United States and Oliver Heaviside in England, investigated why a wave should travel such a curved path. Almost simultaneously they found that there is a reflecting layer in the atmosphere high above the surface of the earth. This layer, now called Kennelly-Heaviside layer or E-region, consists of charged particles at an altitude of 80 to 160 km (50 to 100 miles) and acts like a mirror for certain wavelengths of electromagnetic radiation.

But, despite this fact, there is still serious doubt that Marconi and his assistant actually heard a signal from Poldhu in 1901. As was later discovered, this E-layer works much better at night, especially for the frequency Marconi used (about 800 kHz). The time of transmission was the worst possible choice, when the entire 3470-km (2200-mile) signal path was in daylight [6]. Three months later, on a ship traveling from Southampton to New York, with a better receiving antenna and a more sensitive receiver, Marconi was able to receive Poldhu at a distance of up to 3350 km (2100 miles) at night, but only 1120 km (700 miles) during daylight [23].

Did Marconi then lie about receiving the signal? There is no reason to believe that he did; he was fully aware that, if it later turned out that such a transmission was impossible, it would have utterly ruined his reputation. But put yourself in his place on December 12. 1901: you are in desperate need to prove an idea. The wind is howling outside and you have been listening to the crackling and crashing in your earphones for an hour and a half and, most importantly, you know what you expect to hear: three clicks. Marconi's situation is very similar to staring at the screen of a TV set tuned to an empty channel. If you stare at the noise long enough you can see any pattern you want. Marconi's experiment was badly designed: he knew what to expect; such an experiment would not be considered scientific today.

It took a long time after December 12. 1901 for transatlantic wireless communication to become practical, but the announcement gave Marconi an enormous advantage. Although still far from profitable, he moved energetically to gain a monopoly in the field. Marconi patented everything and what he couldn't patent himself he bought up. His drive, perseverance and ability to hire competent people made him the early leader in wireless and his company the dominating manufacturer and operator of wireless stations. As he became famous and then rich he began to crave the company of the social elite. But despite his outward brashness, Marconi was plagued by self-doubts. In private he was often nervous and irritable and sometimes became so melancholy that he lost the desire to live, then suddenly snapped out of it and worked long hours. He bought himself a 73-meter (220 feet) yacht, the Elettra, and used it both for transmission experiments and to throw parties for the titled and rich [29]. He was made a Senatore, a Marchese, a Citizen of Rome and, by government appointment, President of the Royal Academy of Italy [33]. None of this seemed to make him very happy; his bouts with melancholia increased steadily until he died in 1937.

In 1880 Edison was working on the light bulb, trying to eliminate the blackening on the inside of the glass. It was quite apparent that the filament threw off the deposits. The idea came to him that he could electrically probe the interior of the bulb with a separate wire sealed through the glass, but he was surprised when he found that a galvanometer, connected between this wire and the filament, registered a small current. He didn't know what to make of it [35].

Edison returned to this effect again in 1882 and a year later he applied for a patent on a voltage indicator. He still didn't know what this effect was all about, but at least he had found an application for it.

The man who investigated this "Edison effect" further was John Ambrose Fleming, who was a consultant to the Edison Electric Company in London. He wrote a paper about it, and then put the project aside. Two years later he resumed his investigation, but there was nothing conclusive again. As everyone else, Fleming was still going by the assumption that this current flow was due to atoms or molecules emitted from the filament, the same atoms or molecules which blackened the inside of the glass. He put the project aside again.

Six times over a period of 16 years Fleming took his experimental bulbs out of the cabinet or had some new ones made. Each time he learned a little more about the effect and each time he wrote a paper, but there was never anything conclusive [35].

In 1899 Fleming became technical advisor to the Marconi Company. He knew that finding a better detector was an urgent requirement, yet five years went by before Fleming made the connection between the Edison effect and the need for the detector. The radio signal came in as rapid oscillations, either in packets (for Morse code) or in fluctuating strength (for speech). The oscillations needed to be of high frequency, far above what one could hear. What was needed was a way to smooth them out, leaving only the Morse signal or the audio.

The problem with smoothing out, though, is that the radio signal is symmetrical - the positive halves of the wave are exactly equal to the negative ones - and thus its average is always zero. Fleming now realized that these signals first had to be rectified (i.e. one half had to be eliminated) before it could be smoothed out, and that this was exactly what the Edison effect was capable of doing. In his bulbs the current flowed only from the hot filament (the cathode) to the wire (the anode) but never in reverse and thus it would let only one half of the radio signal pass [16]. In the paper announcing the device Fleming talked about electrons, indicating that he had come to the (correct) conclusion that electrons, not atoms or molecules, constituted the current flow in the valve, a recognition that would become important in a few years [35].

Fleming's invention is the forerunner of the vacuum tube (or valve), but it enjoyed very little commercial success because it was almost immediately replaced by the crystal detector [17], developed by Greenleaf Pickard, an engineer with AT&T, and H.H.C. Dunwoody, a retired Signal Corps general.

Pickard's story is the more interesting one and, with apologies to the general, we shall follow it here. Pickard had tried at one time to eliminate the annoying static in a receiver by reducing the voltage to the coherer, cutting out two of the three batteries. He found an improvement, but later he discovered that he had mistakenly cut out all three batteries, leaving the coherer with no power at all [13].

This intrigued Pickard and he started to experiment in his spare time, listening to the Marconi telegraph station at Wellfleet, Massachusetts, which transmitted regularly between 10 pm and midnight [13]. He believed that in his detector the minute contact area was heated by the radio waves and the heat then caused a (thermoelectric) current. He tried a variety of materials - metals, oxides, sulfides - all with good thermoelectric properties, and all of which happened to have a high resistance.

In 1905 he found a reference to silicon, which also had a high resistance and which, Pickard reasoned, should therefore work, too. But there was no silicon available in Boston and it could be ordered only in powder form. After a long search he found a small quantity of fused silicon at Westinghouse. He tried and it worked better than any other material he had. First as a part-time venture, later full-time, Pickard started the Wireless Specialty Apparatus Company, selling crystal detectors, a venture that turned out to be nicely profitable [13].

Neither Pickard nor Dunwoody knew what exactly the crystal rectifier did in the receiver circuit; all they knew was that it worked - most of the time - and when it worked it worked better than either Fleming's valve, the coherer, or just about any other form of detector. It was not until 1909 that George Washington Pierce, a professor at Harvard, using one of the first laboratory oscilloscopes, determined that the crystal detector was in fact a rectifier and its function and operation was identical to the valve [13]. But how the crystal managed to rectify, nobody knew for sure; this discovery had to wait for science to catch up with invention.

The crystal detector was an odd little device. A piece of crystalline material - silicon, carborundum or galena - was held in a clamp, which formed one terminal. The other terminal was a fine wire, called a cat's whisker, which contacted a small area of the crystal. Some areas worked, others did not, and nobody knew why. The user simply probed around the crystal with the wire until the most sensitive area was found.

Lee de Forest was born in 1873 in Council Bluffs, Iowa, the son of a Congregational minister. When he was six years old, his father accepted the presidency of Talladega College and the family, including an older sister and younger brother, moved to Alabama. Talladega College had been founded by missionaries as a school for Negroes and the de Forest family found itself ostracized. Often the white town boys hurled rocks at the de Forest children, and they retaliated in kind [10].

As a boy Lee had an ungovernable temper that annoyed his strict and severe father and resulted in frequent spankings, during which Lee stubbornly refused to cry. He grew up with few friends and often withdrew, lying on the floor and sketching inventions.

Although his father pressured him to become a minister, Lee was determined to be an inventor. He found a scholarship at Yale (which had been set up by a distant relative for "anyone named de Forest"). His Yale classmates found him cocky, brash and loud. He was voted the "homeliest" and "nerviest" and on the question who was the brightest he only received one vote (his own) but 16 for "thinks he is".

De Forest, with his bushy eyebrows, sunken eyes, prominent cheekbones and peculiarly shaped head, did not like his own appearance. He was graceless and awkward in his contacts with people, always saying the wrong thing at the wrong time [10]. Being desperately poor, he needed to supplement his small allowance with menial jobs. During his undergraduate days he constantly thought up schemes to make money - an airship, an improved pipe, a chainless bicycle, a crease perpetuator - none of which were successful. He promoted souvenir programs for regattas and proms, with the result that his mother had to bail him out of his debts.

Lee de Forest

When he entered a $50,000 contest for the design of an underground trolley cable and heard nothing, he decided he had been robbed and sent a threatening letter to the contest administrators [11].

Despite an undistinguished undergraduate record, de Forest managed to enter graduate school. In the second year of his postgraduate work he heard a lecture on the work of Hertz, which made him decide to choose wireless as his career. He went to New York for an interview with Tesla, but was turned down.

De Forest received his Ph.D. in 1899 and found a job at Western Electric in Chicago, working for $8 per week in the generator department. Two months later he was promoted to the telephone laboratory [28]. His work there was mediocre, he spent much of the company's time on his own wireless projects.

In May 1900 he quit Western Electric and went to work for a Dr. Johnson, who was in the process of starting up the American Wireless Telegraph Company in Milwaukee. Again de Forest worked on his own device and, when he refused to let the company use it, the association came to an abrupt and premature end [28].

Returning to Chicago he went into partnership with Edwin Smythe, a colleague from Western Electric. Smythe contributed $ 5 of his $ 30 weekly salary. De Forest earned another $5 per week from part-time teaching and editing of technical journals. The two of them got permission to use a laboratory of the Armour Institute when no class was in session [26].

What de Forest was working on was not revolutionary. Many people were trying to improve on the coherer and several had come up with more sensitive detectors. De Forest had found an article on a French detector in which a pasty mixture replaced the metal filings; he experimented with the mixture and filed a patent application on essentially the same device. The device never found a practical use [11].

Nevertheless de Forest felt ready for the big time. Professor Clarence Freeman of the Armour Institute had a design for a new transmitter and was taken in as a partner. De Forest went to New York to set up a wireless link for the reporting of the America's Cup races. He found that Marconi had already signed up Associated Press, so he made a deal with Publisher's Press, which was to pay him $800 if the equipment worked. A former mayor of New York advanced $1000 and became a partner, too [26].

Everything went wrong in New York. Freeman's transmitter did not work; de Forest hastily replaced it with an ordinary spark coil. Then Marconi and de Forest found they were interfering with each other, neither of them using any kind of tuning in their receivers. They agreed to transmit alternately for 5 minutes, but then a third transmitter appeared, operated by the American Wireless Telephone and Telegraph Company, and wiped out all radio communication during the races [26].

De Forest was broke. He begged Smythe for some money, pounded Wall Street to raise capital and, finding none, was ready to ask the Marconi Company for a job. But then he found Henry B. Snyder, an over-the-counter stockbroker, who knew how to go about raising money. Snyder got in contact with a certain Abraham White - and de Forest was in business again [26].

Abraham White was a colorful figure, with flaming red hair and moustache, blue eyes, patent leather shoes, gold chain, silk hat and a flower in his buttonhole. Little is known about his background, except that he came from Texas and had changed his name from Schwartz (Black in German) to White. Needless to say, White had made his money speculating [26].

When de Forest met White he was wearing his only pair of trousers, the lining in his suit jacket was gone, his shoes needed soles, he was hungry and owed money for room and board. White peeled off a $100 bill and told de Forest to get some new clothes [26].

The American De Forest Wireless Telegraph Company was now formed, with White as president and de Forest as vice president and scientific director. De Forest built two demonstration stations, one on Staten Island, and the other in a striking, glassed-in penthouse laboratory on the roof of 17 State Street, to which White brought prospective investors. After the demonstration the guests were treated to a sumptuous lunch, during which a surprised de Forest heard White explain how the company would build wireless stations all along the Eastern Seaboard, the Gulf of Mexico, and across the continent, compete with the telegraph and telephone, and have subsidiary companies all over the world. At that time de Forest had achieved a maximum distance of 11 km (7 miles) [26].

Investments came rolling in. At the St. Louis Exposition in 1904 the company erected a 100-meter (300-foot) tower and de Forest lived in a three-story brick house, with carriage and coachman, cook and butler. He now wore a derby and smoked fat cigars; on paper de Forest was worth nearly $1 million [26].

It was mostly a stock selling scheme. By 1905 more than 90 antenna towers had been constructed, few of which were ever put to use. What little real business the company had was in trouble. A contract for five Navy stations was completed late and wireless links for meat packers between Chicago, St. Louis, and Kansas City were so full of static that the meat packers returned to using the telegraph [26].

The bottom began to fall out of the scheme in November 1905, when a court injunction was obtained against the detector de Forest was using. A warrant was issued for his arrest. He decided to flee to Canada until White could arrange for a $5000 bail bond [26].

At this point White decided he no longer needed de Forest. Indeed, de Forest had come up with few inventions of value. White offered de Forest $1000 for all of his stock and threatened to revoke the bail bond if de Forest did not accept. De Forest accepted. White then reorganized the American De Forest Wireless Telegraph Company into the United Wireless Telegraph Company, which went on for a few more years bilking gullible investors [26].

But, although de Forest was in deep trouble during 1905 and 1906, these were his most interesting and crucial years. Back in Chicago he had noticed that a gaslight in the room flickered whenever the spark transmitter was turned on. It turned out to be merely due to the crackling sound made by the spark, but the scene never left de Forest's mind. Early in 1905 he applied for several patents for radio-frequency detectors, each of them employing either a Bunsen burner or an enclosed electric heating element. The patents claimed that radio frequency, applied to two electrodes near these burners or heaters affected the conduction of the gas between them. There is no evidence that any of the devices ever actually worked [37].

De Forest was in desperate need for a patentable detector. He investigated Fleming's valve [37] (later consistently claiming that he had not heard of Fleming's work), did not understand how it worked, but began to concentrate more and more on a filament encapsulated with an additional electrode in a glass bulb. Trying (and patenting) everything that came to his mind, he decided to enclose a second anode and then varied the positions of the two anodes with respect to the filament. He found the device worked best as a detector when one of the "wings" was made of meandering wire and located between the filament and the other anode. He patented this as a "Device for Amplifying Feeble Electrical Currents", though there was actually no amplification. As a detector de Forest's "Audion" was about on par with Fleming's valve and inferior to the crystal detector.

De Forest gave a long and hazy paper on the "Audion" to a meeting of the American Institute of Electrical Engineers in New York. The Audion was partially evacuated; he explained its effect as being due to the remaining air, which underwent "changes when subjected to Hertzian waves" [11].

What had de Forest invented here? His device later became the most important electronic component - the vacuum tube - and its appearance marked the beginning of electronics. But we must not be blinded by what happened later. De Forest did not understand how his Audion worked and up to this point had not achieved any amplification.

The most important property of a vacuum tube is amplification: with a tiny amount of power you can control a much larger one. It is akin to a water valve with which one person can turn on a very large turbine. And the vacuum tube (or valve) can do this gradually, in fine increments.

A device such as the vacuum tube was badly needed and nothing illustrates this better than the quest to transmit voice (or music) over radio.

If you have a vacuum tube you can amplify the feeble signal from a microphone until it is strong enough to control the power going into the transmitter. Thus the transmitter power increases and decreases in rhythm with the voice (called amplitude modulation, or AM). Reginald Fessenden tried to do this in 1900, before the vacuum tube existed. He built a massive, water-cooled microphone that directly modulated the amplitude of the transmitter. It didn't work very well.

During the same time Robert von Lieben in Vienna developed and patented a tube which used a complete vacuum and produced amplification. But he received very little publicity and died of cancer at age 35 [37].

De Forest now repeated the history of his first company: he turned to James Smith, one of White's star salesmen, to raise money for the De Forest Radio Telephone Company. Smith became president, de Forest a vice president.

In 1908 de Forest married; he now received a salary of $300 a month and had seven assistants. With increasing stock sales the value of de Forest's shares shot up to $700,000. He started to build a large house outside New York [26].

The De Forest Radio Telephone Company sold a variety of radio equipment products. In its advertisements the Audion is either not mentioned at all or as a minor item; it simply did not work that well. The crystal detector performed better and was less expensive. Only a few hundred Audions were made per year and used only as detectors. He allowed the British patent on the Audion to lapse for non-payment of a $125 renewal fee [11].

To continue selling stock de Forest now tried two publicity stunts. In the first one he attempted to transmit voice between the Metropolitan Life Insurance Tower in New York, the tallest building in America, and the Eiffel tower, the tallest structure in Europe. He began erecting antennas in 1909. In New York the antenna stood for a year, then a wire snapped and it fell to the ground, narrowly missing a pedestrian. After this incident he was asked to remove the antenna; no transmission between the two points had ever taken place [10].

In the second attempt de Forest set up an opera transmission with Caruso singing. The few people who had receivers found the music so badly distorted that the voices were hardly recognizable [28].

It was Smith who delivered the coup de grace to the company. He calmly informed the board of directors that the last block of 20,000 shares sold were not the company's, but his own, and then walked out, leaving behind a debt of $40,000. De Forest attempted to reorganize the company and its subsidiaries into a new shell, the North American Wireless Corporation, but the government had belatedly started a crusade against radio stock promoters and de Forest and his companies went into bankruptcy. He lost his dream-house, his wife sued for divorce, and he was forced to accept a job with the Federal Telegraph Company in San Francisco [26].

To make matters worse, de Forest, Smith, two other former associates and the stock underwriters were indicted in 1912 on mail fraud, alleging that the four of them had used fraudulent methods to sell stock. Smith was found guilty and drew a heavy fine and jail sentence. De Forest was cleared on three counts, but on the fourth, conspiracy, the jury disagreed. The government decided not to pursue the case any further [37].

Shortly after de Forest started to work at Federal Telegraph, he received a letter from a John Hays Hammond, telling him of the work of a Fritz Loewenstein in New York. Loewenstein had taken some of de Forest's Audions and had found out how to use them properly, biasing the third electrode, the grid, with a small negative voltage and thus allowing it to control the current flow between cathode and anode in a predictable way [37].

It was probably this letter that prompted de Forest to take a second look at his Audion. He managed to build an amplifier and reckoned that this amplifier was just what was needed for the telephone. On October 30, de Forest was able to demonstrate his amplifier before a group of AT&T engineers, with the purpose of selling the Audion. The demonstration was repeated for Dr. Harold Arnold, who had studied under Millikan and knew much about cathode rays and vacuum devices. The amplifier did not work very well, it could handle only weak signals and would distort badly with louder voices and show a blue haze inside [37].

Arnold immediately sensed that de Forest's explanation of how the Audion worked was wrong. He had enough experience with electrons in vacuum to know that the residual gas hindered the desired effect rather than helping it.

AT&T took its time. Not only needed a lot of work to be done to determine if the Audion could be useful for the telephone system, but de Forest had assigned his patents to his old, bankrupt company and a bewildering number of subsidiaries, which needed to be tracked down.

Finally, in 1913, de Forest received $50,000 from AT&T for the use of the Audion in all fields except wireless telegraphy and telephony. De Forest used some of the money to buy back his dream house. A year later AT&T paid an additional $90,000 for a non-exclusive license for the use of the Audion in wireless telegraphy [37].

Arnold and his engineers at AT&T worked hard to improve the Audion and a string of minor improvement patents resulted. De Forest noticed the issuance of these patents and filed patents with identical claims for the sole purpose of interference [12]. It was an incredibly underhanded action (which he blithely admits in his autobiography, immodestly entitled "Father of Radio"), but it worked. In 1917 AT&T made a final payment of $250,000 for all remaining rights, except sales to direct users, the U.S. Government, a license to Marconi, and the distribution and reception of news and music over radio [37].

With the payments from AT&T, de Forest was flush with cash for the first time in his life. But his troubles persisted. The Marconi Company, owner of Fleming's patents, sued for infringement in 1914. De Forest counter-sued, claiming that the Audion was not based on Fleming's valve and that he had not known of Fleming's work when he invented the Audion (a statement clearly untrue). De Forest lost in court and again on appeal [37].

He spent his money lavishly on his house, his laboratory, and making talking pictures. By 1936 he was bankrupt again - and divorced three times [26].

After this he moved his base of operation to Hollywood and, at 57, he married a 21-year-old starlet; strangely enough it was the only marriage that lasted. The remaining two decades of his life he spent lecturing, consulting, and writing, never becoming financially successful again. He died in 1961, at age 87, his estate having dwindled to a little over $1,000.

David Sarnoff was born in Russia, in a small Jewish village near Minsk. His father emigrated to the United States and, in 1900, the rest of the family followed. At first they lived on New York's East Side, later in a section aptly called "Hell's Kitchen". David grew up a tough, streetwise boy, selling newspapers to supplement the meager family income. When he was 17 years old his father became an invalid; David quit school and found a job, first as messenger boy for a telegraph company and then as office boy for the Marconi Wireless Telegraph Company. By learning Morse code in his spare time he advanced to radio operator. This position allowed him to move around among the various Marconi stations and spend considerable time on ships [28].

One of the Marconi installations, at Wanamakers department store, was a powerful 5-kilowatt station ostensibly set up to communicate between the stores in New York and Philadelphia; but it was really more a publicity gimmick. Sarnoff served as Marconi operator at Wanamakers in 1912. On April 12 he received a message from the S.S. Olympic: "S.S. Titanic ran into iceberg. Sinking fast." Sarnoff notified the press and then stayed on duty for three days and three nights, copying down the names of survivors and handing them to the press [26]. Or so the story goes. Sarnoff actually made part of it up and embellished the rest [25]. When it came to image building, he never had much use for the truth.

The Titanic disaster drastically demonstrated the importance of radio and how it had been widely misjudged. The captain of the Titanic had ignored several messages of iceberg sightings in its path, transmitted only hours before by other ships; the ship nearest the Titanic did not respond to the SOS because there was no radio operator on duty at night. Within months almost all maritime countries passed laws requiring that ships with more than 50 passengers be equipped with wireless sets and that an operator be on duty day and night [26].

After this publicity, David Sarnoff rose rapidly through the ranks of the Marconi Company. By the end of 1912 he was radio inspector for ships in New York harbor, a few months later chief inspector for the entire country. In 1914 he was made contract manager, a newly-created position which linked engineering and sales and for which Sarnoff was the ideal man, having acquired an understanding of the technology and being a natural-born salesman. In January 1917 he became commercial manager, the third highest management position in the company; he was 25 years old [26].

Sarnoff was a visionary. In 1915 he wrote a memorandum to the higher-ups at Marconi, proposing that the company build radio music boxes - receivers for the ordinary household - and that stations be built solely for entertainment. The memo was ignored.

At the end of World War I, the United States military became critical of Marconi. This "foreign-owned" company controlled almost the entire field of radio communication. In 1919 Marconi wanted to buy 24 high frequency alternators from General Electric and negotiated for exclusive rights. It was a large, lucrative contract but the man in charge of the negotiation at GE, Owen D. Young, felt it was necessary to inform the Navy. The acting secretary of the Navy, Franklin D. Roosevelt, delegated an admiral, who asked GE not to issue an exclusive license but that instead GE propose to buy the American Marconi Company. The negotiations were delicate, but with the power of the U.S. Government behind GE and the threat of loosing this major customer anyway, Marconi agreed to sell.

David Sarnoff

GE named its subsidiary the Radio Corporation of America. Young was made chairman, and Sarnoff at first commercial manager, then president. It was intended to be mainly a selling organization, with GE doing the manufacturing. In 1920 AT&T bought a large block of RCA shares from GE and a royalty-free cross licensing agreement was signed between RCA, GE, and AT&T [27]. Westinghouse, realizing that as a late starter it was being left out of a good thing, began to buy up radio patents, thus forcing itself into the new combine. In 1921 the cross-licensing agreement was extended to Westinghouse; GE was to manufacture 60 percent of the equipment, Westinghouse 40 percent [27].

It all might have worked out if the radio business had remained what it was then: the building of stations for the government and a few large ship-owners. The acquisition of the American Marconi Company and the subsequent cross-licensing agreements were clearly against the anti-trust laws -- which might have been justifiable for national security. But the radio business took a drastic turn, a turn that Sarnoff had foreseen in his 1915 memorandum, but for which neither RCA, GE, nor Westinghouse were prepared. Amateur wireless enthusiasts started building receivers - mostly crystal detector sets - first for themselves, then for friends and neighbors, and a whole population began to listen to anything that was on the air. There was a need for commercial stations and for vacuum tube receivers, which independent entrepreneurs were eager to manufacture without worrying too much about the patents controlled by RCA. At the end of 1922 there were 30 licensed broadcasting stations in the United States; by 1924 over 500 [27].

RCA controlled almost all radio patents - by now almost 4000 of them - but it was tied to the lumbering factories of GE and Westinghouse and could not keep up with the rapidly moving independents (such as Atwater-Kent, Grigsby-Grunow, Crosley, Philco, Zenith and Emerson) who came out with improved models every year. So RCA began to enforce its patents, trying to squeeze the independents out of business or, if that was not practical, to extract stiff royalty fees. The reaction was not long in coming: the way RCA had been set up violated the antitrust laws and in 1930 the Government sued. In 1932 a consent decree was signed, requiring AT&T, GE and Westinghouse to dispose of their holdings in RCA, for GE and Westinghouse not to manufacture radio sets for 2 1/2 years and for all cross-license agreements to become non-exclusive [27].

The intent of the Government action was to break up a monopoly, but RCA emerged from the debacle stronger than ever. It acquired the GE and Westinghouse radio set factories and Sarnoff, no longer tied to the slow decision-making processes of his suppliers, was now in a much better competitive situation. RCA still controlled almost all radio patents [27].

Now the time has come to meet the real hero of radio: E. Howard Armstrong. He did not have the executive drive of Sarnoff, he was not a promoter like de Forest, nor did he have the flamboyance of Marconi, but it was Armstrong who made radio possible.

He was born in 1890; When he was 12 years old the family moved to a large, comfortable house in Yonkers. There Armstrong began to experiment with electricity and radio in a large attic room. This hobby became so important to him that the family nicknamed him Buzz, from the buzzing sound of his spark transmitters [24].

After high school he was admitted to the Columbia School of Engineering, to which he could commute from Yonkers on the red Indian motorcycle he received as a graduation present. Now 19 he was a lanky, modest and somewhat retiring young man who had the air of knowing exactly where he was going. In his junior year in college he met Professor Michael Pupin, the Serbian immigrant who had already contributed considerably to the state-of-the-art of telephone transmission.

With Pupin's encouragement Armstrong became interested in the Audion and soon felt strongly that de Forest had misinterpreted his triode. He made an exhaustive study of the characteristics of the device. In the summer of **1912**, before his senior year, he was suddenly struck by an idea: what would happen if he let the tube amplify the radio signal, use a tuned circuit to filter out other stations and then feed the enlarged and purified output back to the grid to be amplified again and again? After vacation he tried it and it worked. Within weeks he had this "regenerative" circuit working astonishingly well. The improvement in his receiver was stunning; the signals were so loud that the earphones could be left on the table and he was able to listen to several stations in Europe, as well as San Francisco and Honolulu [24].

Armstrong wanted to patent the idea immediately, but his father refused to advance the $150 filing fee, feeling that Howard had spent too much time fooling around with his radio equipment and too little on homework; he would pay for the patent application only after Howard had graduated. So the best Armstrong could do was to get the schematic notarized [24].

There was another invention hidden in his regenerative circuit. When he tried to achieve too high an amplification, i.e. when he fed too much of the amplified signal back into the input, the circuit began to howl. What happened was that the circuit became an oscillator, generating its own, sustained frequency, and could thus be used as a transmitter.

Armstrong graduated in June of 1913 and became an assistant at Columbia at $600 per year. The patent application for the regenerative receiver was filed in October and for the oscillator in December [24].

In the fall of 1913, de Forest gave a paper to a meeting of the Institute of Radio Engineers (IRE - later merged with the AIEE and renamed IEEE) at Columbia. Afterwards Armstrong demonstrated his regenerative receiver to de Forest, but wisely kept the equipment hidden in a box and did not divulge how it worked. There was instant dislike between the two men. The fact that Armstrong, 22 years old and a recent graduate, was sure that he had something better than de Forest, 40 and a Ph.D., did not help.

In December, a delegation of the American Marconi Company, including David Sarnoff, came to Pupin's laboratory and listened to Armstrong's receiver. Sarnoff was impressed and recommended to the company management that Armstrong's patent be acquired. But the Managing Director of the parent company in England thought it was a rather wild idea and turned the request down. In February 1914 AT&T engineers came by; again, there was no sale [24].

The antagonism between Armstrong and de Forest developed into an open clash in the fall of 1914, after Armstrong published a paper on the "Operating Features of the Audion" and gave a paper to an IRE meeting early in 1915 [34]. He correctly interpreted the mechanism of the tube as a stream of electrons emitted by the hot cathode, collected by the anode (or the plate) and controlled by the grid in between the two. De Forest attacked Armstrong's findings as erroneous, still believing that it was the gas in the tube which somehow made it work. When Armstrong described his oscillator, de Forest made the statement that the oscillating Audion was based on some other action [24].

While this confrontation was taking place, de Forest took two surprising actions. In March 1914, six months after his visit to the Columbia University laboratory, he filed for a patent on a circuit he called the "Ultra Audion", which, according to de Forest, would do the same as the regenerative circuit without employing feedback. Indeed, there was no tuned circuit, but in fact de Forest's invention employed exactly the same principles as Armstrong's regenerative circuit [24], but had a poorer and far less predictable performance.

In September of 1915, at the very time he was ridiculing Armstrong's work at professional meetings, and one year after Armstrong had sent in his application, de Forest filed for a second patent, using Armstrong's oscillator circuit [24].

In the summer of 1915 Armstrong's father died of a stroke. He was now the head of the family and had to depend on royalty income to supplement his meager salary. The royalties gradually rose as more and more people realized that the regenerative circuit was the only way to make a sensitive receiver; by 1917 payments reached $500 per month [24].

When the United States entered the war, Armstrong enlisted in the Army. He was now 27 and, flushed with his accomplishments, had lost some of his shyness. After a brief period of training he was made a captain and stationed in Paris as an expert in radio communications. The state of the Army radio equipment was poor, none of it used any vacuum tubes. Armstrong and his crew introduced improvements as fast as they could. Up to this point almost all radio communication took place at very low frequencies, mostly below today's AM band (i.e. less than 800kHz); early vacuum tubes simply could not handle higher frequencies. Rumor was that the Germans were using high frequencies (as high as 3MHz!), a rumor which, after the war, was found to be false, but nevertheless got Armstrong thinking of ways high frequencies could be received [24].

There were bombing raids by German planes over Paris and Armstrong often went outside to observe. It was not a very dangerous thing to do; the planes consistently missed their targets. But even worse was the accuracy of the anti-aircraft guns. Armstrong mulled over the possibility of guiding the anti-aircraft guns by the high frequencies emitted from the ignition system of the plane's engine, when he was suddenly struck by an idea. The high frequency - any high frequency - could be received by mixing it down to a frequency which was within the amplification capability of the vacuum tube [24].

He immediately started to build such a receiver, calling it the superheterodyne. It worked beyond expectation, providing an amplification of several thousand with unprecedented stability, selectivity and quietness. But the war ended before it could be utilized to direct the guns or detect the nonexistent German high frequency communications. Armstrong now had a new invention which, by the U.S. Army rules in effect at that time, he was allowed to keep. Still on active duty in Paris (and now a major), he filed for a patent in February 1919 [24].

The superheterodyne was to eclipse the regenerative circuit in importance; even today almost all radio and television sets employ this principle, no longer because tubes (or transistors) are incapable of reaching the high frequencies, but because it provides outstanding performance. In Armstrong's invention the high frequency (such as 100 MHz) is mixed with a frequency of an oscillator in the receiver, tuned to say 90 MHz. Among the artifacts of the mixer is the difference between the two frequencies, 10 MHz, which not only is a lower frequency but also remains constant as the receiver is tuned to different stations and is thus far less cumbersome to filter and amplify.

In February 1919 the Institute of Radio Engineers awarded Armstrong its Medal of Honor for the invention of the feedback circuits.

Returning from military duty he went back to his old job at Columbia University and found that his second patent applications (the oscillator) was tied up in interference with de Forest's claims. Undaunted he worked to improve the superheterodyne, reducing the many tuning knobs to two to make it easier to operate [19]. Westinghouse, which was now on its patent acquisition spree to maneuver itself into the RCA/GE/AT&T combine, expressed interest in buying all of Armstrong's patents. On October 5, 1920, he signed over his patents to Westinghouse for $ 335,000, payable over 10 years. An additional $ 200,000 were to be paid when and if the oscillator patent was issued [24].

Howard Armstrong

Financially, Armstrong's future was now secure, but his problems were just starting. The oscillator patent continued to be held up by de Forest's interference proceedings, while the De Forest Radio and Telegraph Company was using the regenerative circuit without paying a license fee. Armstrong, backed by the Westinghouse legal department, sued de Forest for patent infringement [24].

While this was going on Armstrong came up with invention number four. He was preparing his regenerative circuit one day in 1921 for a demonstration in court when he was surprised to find a reception much better than usual. Normally the amount of feedback was increased by the listener to the point just before it became too much and began to squeal. In this circuit the feedback needed no adjustment and no annoying squealing was heard. He named it the super-regenerative [24].

Within weeks the super-regenerative became the hottest item in the booming radio market. It required fewer vacuum tubes than the superheterodyne, which was important when tubes were still quite expensive. He demonstrated the circuit to Sarnoff, who decided to buy it. In June 1922 RCA acquired the patents for $200,000 cash and 60,000 shares of RCA stock. When the circuit went into production Armstrong was called in to iron out last-minute problems and received another 20,000 shares. At $38 a share Armstrong was now worth some $3 million.

On top of the world, he went to Europe and brought back the king of cars, a $11,000 Hispano-Suiza. The following year he married Sarnoff's secretary and took up residence in a high-rise overlooking the Hudson at 86th Street and Riverside drive. At Columbia he was elevated to assistant professor. He taught no classes but gave occasional lectures and spent his time on research - and his fight with de Forest [24].

There were two separate legal actions going on simultaneously, the interference proceeding started by de Forest and the patent infringement suit initiated by Armstrong. De Forest claimed that he had invented the feedback principle in 1912 while he was working on the telephone amplifier in the laboratory of the Federal Telegraph Company. And indeed his laboratory notebook showed an entry where he and his assistant had been bothered by a howling sound produced by the amplifier. He did not understand it, found it a nuisance, and did not pursue the effect or make a move to patent it until long after Armstrong had demonstrated his circuit. A comment on the pertinent page in the notebook reads: "Found this arrangement no good for a two-way repeater." [24, 37].

It was an absurd legal situation. Both men had assigned their patents to large companies. De Forest was backed by AT&T, Armstrong by Westinghouse, two companies now linked by cross licensing agreements. But there was also a sinister aspect: if de Forest won, the RCA-AT&T-GE-

Westinghouse trust would have a newly-issued patent with a lifetime of 17 years; Armstrong's patent was already ten years old [24].

Armstrong won the patent infringement suit. De Forest appealed and lost again. But then the Special Master of the Court, appointed to assess damages, died, and no final judgment could be entered until a new one was appointed. De Forest, staring bankruptcy in the face at that moment, would have been glad to call an end to the legal proceedings. But Armstrong would not allow Westinghouse to have the court enter the judgment by waiving damages [24].

It was Armstrong's big mistake. He wanted the cheat brought to justice, to put him out of business once and for all. But it did not work out that way. De Forest retaliated with legal maneuvers in the Court of Appeals in the District of Columbia, pressing his interference case. And in a first-class piece of legal obfuscation his attorneys managed to have the court declare that his telephone amplifier and the regenerative circuit were one and the same. The patent office now had no choice but to issue a patent to de Forest, which meant that two people held a patent for the same invention [24].

Inevitably, more legal actions followed: a suit in the District Court of Delaware against de Forest and another suit in Philadelphia by de Forest against Westinghouse to have Armstrong's patent declared null and void. De Forest won; the courts relied on the decision of the District of Columbia Court of Appeals. Enraged, Armstrong brought his case to the Supreme Court, which agreed to hear it in the fall of 1928. Unable to understand the technical matters, the supreme court went along with the lower courts [24].

Armstrong was not ready to give up, but he was at the end of the legal appeals system. The only chance was to bring to trial a different case. In 1931 a small company in Brooklyn was being sued by RCA for infringement of the de Forest patent - for which RCA had administrative rights - and Armstrong took over the defense. He lost the first round, but won on appeal. De Forest, backed by RCA and AT&T went to the Supreme Court in 1934, which again found for de Forest. Michael Pupin and several others raised a hubbub in the press over the garbling of scientific facts by the court, which caused the Supreme Court to change the wording of the decision but not the decision itself [24].

A week after the final Supreme Court decision, at the annual convention of the Institute of Radio Engineers, Armstrong made a move to return his Medal of Honor, given to him 15 years earlier for the invention of the regenerative circuit. The board of the IRE, half of whose members were employed by AT&T and RCA, refused and the members affirmed the decision with a standing ovation [24].

In the midst of his legal fights Armstrong came up with his fifth invention. With Pupin he had been working on the elimination of static. In an age of million-watt transmitters it is hard to appreciate the problem that static interference represented in the early days of radio. It was a very stubborn problem; the crashing noises, mostly caused by atmospheric discharges, had exactly the properties of amplitude modulation. Moreover the power of a lightning bolt was far greater than that of any AM transmitter. The only solution appeared to be to make the receiver very selective, to let in as small a frequency band as possible, at the expense of sound quality.

There were other methods of modulating a radio frequency, notably frequency modulation. Indeed, varying the frequency of the transmitter in the rhythm of speech or music instead of its amplitude had been tried several times, with very disappointing results. An authority on the subject from AT&T, John Carson, gave the coup-de-grace to the subject of FM in a paper to the IRE with the conclusion that FM "inherently distorts without any compensating advantages whatsoever". "Static, like the poor", he declared, "will always be with us." [24].

Having spent some five years on the investigation of the elimination of static, Armstrong and Pupin gave up in 1922. Unsatisfied, Armstrong decided to take another look at FM, despite the burial pronouncements by Carson. At first it appeared that Carson was right, FM did not seem any better to circumvent static. But slowly, over a period of eight years, Armstrong began to grasp that there had been a basis misconception about FM. With the experience in AM, it had been assumed that the best results would be obtained by making the frequency band as narrow as possible. What Armstrong began to realize was that FM would show an advantage only if it was modulated over a wide frequency range and, if this was done, it would be inherently immune to static interference.

Between 1930 and 1933 Armstrong filed four patent applications. Shortly before Christmas 1933 he invited Sarnoff for a demonstration. Sarnoff, who had often stated that he was waiting for someone to come along with a little black box to eliminate static, was impressed by the performance, but did not quite like what he saw: the "little black box" filled two rooms. Worse, FM was an approach entirely incompatible with the already existing AM transmitters and receivers. Nevertheless, Sarnoff agreed to give FM a test and Armstrong's equipment was moved to the top of the Empire State building, where an experimental transmitter and antenna for television had already been installed. But, despite successful tests, RCA made no move to buy the invention; by now Sarnoff had decided that the next big thing to come was TV. After a year Armstrong was politely asked to remove his equipment [24].

Armstrong was now resolved to make FM successful on his own. In November 1935 he went public with his invention in a paper and demonstration to the IRE in New York. As the receiver was tuned, there was a sudden, eerie silence, as if the equipment had been turned off. Then the audience heard a voice announce: "This is amateur station W2AG at Yonkers, New York, operating on frequency modulation at 2.5 meters." A glass of water was then poured before the microphone, a piece of paper was crumpled and torn and an oriental gong was softly struck. The fidelity of the transmission was stunning, and there was no static [24].

Besides the breathtaking clarity, there was another accomplishment: 2.5 meters - 110 MHz. Such a high frequency had never been used before; in fact it was widely believed that the higher the frequency the more limited would be its range. During the earlier tests from the Empire State Building (at 44 MHz) Armstrong had found - as he had suspected - that transmission with FM at high frequencies could even outdistance AM frequencies [24].

Next Armstrong built a 50-kilowatt experimental station at Alpine, New Jersey and petitioned the Federal Communications Commission to assign frequencies for five FM channels. He continued his development work, received 12 additional patents, and promoted FM in lecture tours. There was a great deal of prejudice to be overcome; despite the demonstrations, the conviction that high frequencies could not reach beyond the horizon persisted and many engineers felt that the wide frequency (200kHz) band required by FM was wasteful compared to the AM bandwidth (10kHz). RCA derided the whole idea, stating that the public had no interest in high fidelity [24].

But the idea caught on. In the fall of 1939 there were 150 applications for FM stations on file with the FCC. The following year the commission converted the frequency band reserved for television channel 1 (44-50MHz) into 40 FM channels [24].

During World War II, however, all FM receiver manufacturing stopped. Armstrong, who had so far received only about $ 500,000 in royalties on his FM patents but had spent over $1 million to promote his system, continued with his cause even after all royalties dried up [24].

After the war a fierce battle over frequency assignments developed in the chambers of the FCC. In RCA's view FM was competing with television both for the consumer dollar and the frequency spectrum and the methods the company used to assure clear sailing for television were far from clean. The FCC held hearings and distributed a report which stated that the frequency range then used for FM would soon be unusable because of expected sun-spot activities. A group of radio experts pointed out that the report was in error, but the FCC, clearly under the influence of RCA behind the scenes, abruptly reassigned FM to a new frequency band, 88-108 MHz. Suddenly 500,000 FM receivers became obsolete; it was a disastrous blow for Armstrong and FM. To make matters worse, the FCC adopted a plan to limit the power of FM transmitters to a very low level [24].

Despite this setback, the public wanted the high fidelity of FM and RCA, belatedly recognizing that the television and FM markets would grow simultaneously, started to manufacture FM receivers -- without paying any royalty to Armstrong [24].

For Armstrong it was a desperate situation. He had managed his fortune well, selling much of his RCA stock just before the crash. But the protracted lawsuit with de Forest and the heavy development and promotion expenses for FM had taken their toll. He needed a substantial royalty income, yet his basic FM patents would run out in 1950. So, in 1948, Armstrong sued RCA for patent infringement. If he won the lawsuit, Armstrong could collect triple damages for the entire life of each patent [24].

In February of 1949 the taking of depositions began. Depositions, designed to speed up the proceedings and save the court's time, are taken in a lawyer's office and both parties can question the witnesses. In Armstrong's case they did anything but speed up the proceedings. RCA was playing for time, waiting for the patents to run out. Armstrong was kept on the witness stand for an entire year [24].

As the case dragged on year after year, Armstrong's financial situation became desperate. After 1950 royalties dropped to a trickle while lawyer's bills mounted. In 1953 RCA offered to settle out of court with a $ 2 million "option", but it wasn't very clear just how much money Armstrong would actually be able to collect. In November 1953 his wife urged him to accept the settlement; for some time it had been her wish to retire to a Connecticut farm.

They had a fight and she moved out. He spent Christmas and the month of January secluded in his 13th story apartment. During the night of January 31 1954 Howard Armstrong removed an air-conditioner and jumped out of the opening. A maintenance worker found his body the next morning on the roof of a third floor extension.

Much Ado About *Almost* Nothing

9. Farnsworth

The 15-year old Idaho farm-boy who invented television
and the ruthless self-promoter again.

Paul Nipkow had studied engineering in Berlin. In 1883, when he was 23 years old, he came up with the idea of transmitting images over a wire. He would drill a spiral of holes into a disc. As the disc rotated, the uppermost hole would move across the image, making a thin, bowed strip visible. Then the next hole, located a little lower, would take over, followed by the third, etc. until the entire image was covered and the top hole appeared again. In this way he could scan the image the way you are reading this book, line by line. Behind the holes he intended to have a large selenium cell that would convert the fluctuating light intensity into an equally fluctuating electrical voltage [22].

At the receiving end he would have an identical disc, spinning in step with the first one. A light was to be modulated by an effect that Faraday had discovered: a piece of flint glass rotates the plane of polarization of light if placed in a magnetic field. Nipkow intended to wind a coil of wire around the periphery of the glass and feed it with the signal coming from the selenium cell. The observer then could look through the holes of the disc into the modulated light and see the reconstituted picture [22].

Nipkow got a patent on his invention in 1885 and wrote articles about it, but he never made it work. There was no way he could have made it work, the signal produced by the selenium cell was far too weak to drive the coil around the flint glass and, of course, there were no amplifiers yet.

After he received his degree he joined a company making railroad signals and stayed there during his entire professional career. He let the patent lapse, married, had six children, made a mediocre living and never again came up with any great ideas.

In 1928 he found himself standing in line as one of the crowd at an exhibition where a German company showed a primitive television receiver, using his spinning disc. There, in a darkened room, he saw a flickering, blurred picture. Thus, he witnessed his own invention. In 1935 the Nazis, recognizing the propaganda value of a German inventor of television, celebrated his 75th birthday in grand fashion. He received an honorary doctorate from the Johann Wolfgang Goethe University in Frankfurt and a pension. After his death in 1940 a state funeral was arranged.

After Nipkow's publication there were many schemes proposed, most of them using his basic idea. But none of them worked, they all lacked one ingredient: the amplifier. The vacuum tube appeared in 1906, but it was not until after World War I that would-be television inventors began to notice it. During the same time Geitel in Germany, using potassium-hydride, developed a new and considerably faster light sensitive cell [24].

Just who combined all these elements into a working television apparatus for the first time is still open to speculation.

It could have been Denes von Mihaly in Budapest, Hungary, who had something working in July 1919. But his picture, it seems, was so blurred and coarse that it was quite impossible to recognize detail.

Or it could have been August Karolus, who, in the summer of 1924, was an assistant professor in the physics department of the University of Leipzig. Karolus started his work by developing a better light valve for the receiver, using as the basis an effect discovered by Kerr in which some liquids change their transparency with an applied voltage. He used a Nipkow disc nearly a meter across, with 48 holes (segmenting the image into 48 lines), and a vacuum tube amplifier. The resulting picture was said to be "quite good", but Karolus published so little that we are not really certain what he achieved.

Even though he may not have been the first, the most interesting of these early pioneers was John Logie Baird. Baird was born in 1888 in Scotland and went to the Royal Technical College in Glasgow. He did not do well in school and it took him five years to become an associate [17]. Then he went to Glasgow University, where he took courses in electrical engineering and physics but didn't graduate. In 1914 he became superintendent engineer at the Clyde Valley Electric Power Company.

Baird was a hypochondriac. He was pale, thin and nervous and got ill whenever he was under stress. He hated his job because it involved supervising the repair of power lines at odd times of day or night in all weather; consequently he got sick a lot. Above all, he always complained of cold feet. So in 1918 he invented the "Baird Patented Under-Sock", a regular sock which he sprinkled with borax and sold the to retail stores. He did reasonably well, making some £ 1,600 off the idea. He added "Osmo Boot Polish" to his line but became ill and had to give up the business. What he needed was a warmer climate, he decided, and went to Trinidad, taking along a shipment of cotton goods to sell to the natives; but there was no market for cotton goods in Trinidad. He tried to start a business there making jam, but the insects made off with most of the sugar. Getting ill with malaria and running out of money, he went back to England. While he was away, his girlfriend had married; undisturbed by social customs; she shared her time between her husband and Baird for the next ten years.

Baird now started a business buying Australian honey wholesale at the dock and selling it by mail and then, for £ 100, bought into a business selling fertilizer. But this second business came with a damp office underneath a railroad bridge, so Baird had another health breakdown. He sold his interest and went to the Buxton Hydro to recuperate.

The latter part of 1921 found him in London again. With his last £ 100 he bought into a business that sold soap through travelers. A year latter he had another breakdown, this time a serious one. He sold out to his partner, getting £ 200, and went to Hastings to recover.

Baird was now 35, still in poor health and nearly broke. To make matters worse, his girl friend decided to come and live with him at Hastings for a spell. He needed to make some money in a hurry, so he decided to invent something.

His first idea for a product was a glass safety razor, which would not rust; when he tried it he cut himself badly [18]. Pneumatic shoes were the second idea. He put inflated balloons into large boots, but when tried them out, one of the balloons burst and that was the end of idea number two.

The third idea was television. In his attic room above an artificial flower shop at Hastings he started to experiment, using an old tea chest, a biscuit box, darning needles, wood scrap, some secondhand vacuum tubes, a bicycle lamp lens and a used motor. The Nipkow disc he cut out of cardboard. By early 1924 he was able to transmit the image of a Maltese cross from one end of the table to the other.

He was out of money now and advertised in The Times for investors. There was only one taker: a London cinematograph operator bought a one-third interest for £ 200.

In August 1924 Baird moved back to London and set up a laboratory at 22 Frith Street, Soho. He started to improve his apparatus, developing his own photocell. By April 1925 he needed money so badly that he agreed to give personal demonstrations of his set at Selfridge's department store for three weeks at £ 25 per week. His picture quality at this point was very poor, probably much poorer than that achieved by Karolus the year before. Baird only had some 400 picture elements (18) (also called pixels, i.e. distinct, resolvable points in an image; modern TV set has in excess of 350,000 pixels) and could only transmit either black or white areas, there was no gray scale. A human face appeared as an oval of bright light; when the mouth opened a flickering black spot appeared in the center. Also, he had thus far not dealt with the problem of synchronizing the sending and receiving discs; he simply ran them from the same motor, separated by a long shaft.

The £ 75 from Selfridge's didn't last very long. Soon he was broke again and literally starving. He advertised for a company promoter to raise money. The following day his laboratory was deluged by men in dilapidated white spats and frayed striped trousers, all willing to take the job but all asking for a cash advance [18].

Next he sent letters to 3,000 doctors and dentists. He received seven replies, but no investment [27]. He appealed to relatives, was able to raise £ 500 and formed a company, Television Ltd.

By October 1925 he was able to produce some gray. Captain Oliver George Hutchinson, a former competitor in the soap business, joined Television Ltd. Hutchinson was a promoter and his job was to raise money.

On January 26. 1926 Baird showed his television apparatus to some 40 members of the Royal Institution who came to see him at 22 Frith Street. What they saw was a small (3.8 by 5 centimeter, or 1.5 by 2 inches), flickering and rather blurred picture, using 30 scanning lines and five images per second. According to one observer it was "just possible to recognize faces" [6].

In February Television Ltd. moved into larger quarters. Rather than develop what he had into a marketable product, Baird now aimed at epoch-making breakthroughs, which had little commercial value but were good for publicity, allowing Television Ltd to raise more and more money.

In December 1926 Baird announced infrared television. He had discovered that his photocell picked up not only the visible but also the infrared spectrum (heat). Thus, with a suitable filter in front of the lights, his television receiver could see in the dark. Unfortunately the range was very limited.

In May 1927 he transmitted a low definition picture via telephone line from Glasgow to London. In February 1928 he managed to send a television image via an amateur transmitter from London to New York. In

July 1928 he demonstrated color television, using three sets of holes on each Nipkow disc. On the sending side the three sets were covered by red, blue and green filters. The receiver had three different light sources. Then he topped this by storing a (barely recognizable) image on a phonograph record.

With the publicity created by Baird's experimental blitz, Television Ltd. was able to raise some £ 250,000 from the public and, for a brief period, Baird was rich. He started to live well, but his poor constitution didn't agree with the new lifestyle and he got sick again, repeatedly [18].

Television Ltd. had another very serious problem. Public communication was controlled by the post office and the BBC had a monopoly. Baird and Hutchinson needed the permission and cooperation of both to bring their TV system to the public and they botched the job badly, misstating facts on applications and infuriating the officials with their unscrupulous methods [6]. It was not until September 1929 that the differences were settled and an official trial was made over BBC transmitters. But only one (AM) channel was available and the picture and sound had to sent alternately. First the picture was seen for two minutes, then the speaker repeated what he had said over radio. The quality of the picture was still very poor, Baird had problems with up and down movement of the image, the picture flickered like an early cinematograph and was generally blurred. However, most people were impressed - some commented that they could even see the lips move [18].

In February 1930 the first television receiver was offered to the public for 25 guineas, with a tiny viewing area and a magnifying glass; few were bought. Rather than to improve the picture quality, Baird immediately went off and developed a large screen display using an array of light bulbs. He then arranged three such screens side-by-side for panoramic effect.

The BBC started regular weekly television broadcasts in October 1931. People could only be televised in close-ups and they needed special makeup: a dead-white face, blue lips and blue-shaded eyes. Baird was still using only 30 lines and 12.5 images per second.

As we shall see later in this chapter, electronic television was already well on its way by this time. Baird didn't believe in electronics, his inclination was toward mechanical things. But competition from the new approach soon made itself known. In 1932 EMI (Electric and Musical Industries), had links to RCA and Marconi and started to work on electronic television in Great Britain. By 1935 a 405 line, 50 image-per-second system was operational. Baird, in the meantime, had been able to improve the quality of his images somewhat, but not enough. Between November 1936 and February 1937 the two systems were tested by the BBC, alternating between the two each week. Needless to say, the EMI system won [11]. In 1939 Television Ltd. went into receivership; it had only managed to sell some 5,000 receivers during its existence [6].

Disheartened and rejected Baird continued working on large-screen TV receivers during the war. He died in 1946, about as poor as when he started [19].

That mechanical television was doomed became rather obvious to anybody but Baird. In order to display even a fraction of the detail to which the human eye is accustomed, hundreds of lines needed to be transmitted for each image and, to avoid flicker, at least 25 images needed to be transmitted every second. This called for high speed - thousands of movements per second - and the movements had to be done with great precision. Mechanical gadgets simply had too much inertia to do the job.

It was Ferdinand Braun who gave part of the key to electronic television. He knew that cathode rays caused a glow when they hit a glass wall and that the beam of electrons (and, therefore, the glow) could be made stronger by heating the cathode with a red-hot filament. Braun covered the inside of the glass wall with a phosphorescent layer, so that the cathode ray produced a brilliant spot. Then he added a coil to deflect the beam.

But Braun stopped short of the cathode ray tube (CRT) as we know it today. He was satisfied with beam deflection in one dimension; a second coil was added later to give a two-dimensional display.

In 1922 Justin Tolman, a chemistry teacher, was given a fascinating presentation on a radically new approach to television. The complex diagram on the blackboard and the sophisticated explanation astonished him, especially since he was listening to one of his students, 15-year old Philo Farnsworth.

Farnsworth was a Mormon farm-boy and the place was Rigby, Idaho, in the eastern part of the state. He had read every popular electronics magazine he could get his hands on. He knew about Nipkow and Baird and had studied Einstein on his own by borrowing books from the library.

Einstein had started his work by examining a puzzling effect: in some materials (e.g. selenium) light produced electricity, i.e. an electron can absorb the energy of a light beam to the point were the electron frees itself from the atom. This was the starting point for his famous conclusion that energy can be converted into mass and vice versa, $e=mc^2$.

In an ordinary photocell the electrons freed by light are collected at a contact, i.e. they produce a voltage. (A second terminal allows other electrons to flow back into the material, replacing the ones taken out and thus the effect goes on forever, without piling up a lot of electrons).

Farnsworth took the photocell a step further. In his "image dissector" a flat glass plate was part of a vacuum tube. On the inside, i.e. in the vacuum, the glass plate was coated with a photosensitive material. If a beam of light hit the photosensitive layer, electrons would be released into the vacuum. On the opposite side of the tube there was a plate with a tiny hole in it (an aperture). A strong positive voltage was applied to this plate (the anode), which made the electrons fly in a straight line toward it. Thus the "anode" collected all electrons produced by the light, except the ones moving through the aperture; which were collected by a small metal plate right behind it.

And here comes the part that shows the genius of Farnsworth: He added four metal plates on the periphery, along the path between photo-plate and anode, one on top, one on the bottom, one on the left and one on the right. Applying voltage to these plates he could then *deflect* the electrons as they were traveling. Thus the plate voltages could be set so that the electrons coming through the aperture had originated from the lower left corner, or the upper right one, or any other place on the photo-plate. Thus he could *scan* the image projected onto the photo-plate and convert it into an electrical signal, which then could be transmitted over a wire or a radio wave. An electron beam, having hardly any mass, could be moved much faster than any mechanical contraption and thus he could have more lines and a much higher resolution.

After high school Farnsworth went to Brigham Young University in Provo, Utah. His family was poor, so he had to support himself, which he found difficult to do. He joined the Navy in San Diego and then realized that any invention he worked on during his tour of duty would belong to the government. This prompted him to get a discharge after only a few months. He went back to Utah, found some menial employment, took courses at Brigham Young and, by correspondence, at the University of Utah [7].

In the spring of 1926 he found a temporary job with a fund-raising drive for the Community Chest, headed by two traveling organizers, George Everson and Leslie Gorrell. Farnsworth explained his television ideas to them and estimated that it would take about $ 5,000 to develop a prototype. Everson and Gorrell decided to invest their entire savings of $ 6,000, for which they received a 50 percent stake in the partnership [5].

Everson's and Gorrell's next assignment was in Los Angeles. Farnsworth liked the idea of being near Caltech; he hurriedly got married and moved with his bride Elma (Pem) to California. He commissioned a glass blower to built the image dissector and in their small apartment in

Hollywood he began to assemble his television system. Everson, Gorrell and Pem helped winding coils, making drawings and locating hard to find items [7].

When everything was ready Farnsworth applied power and the whole setup went up in smoke.

There was no more money left, Farnsworth had spent the entire $ 6,000 in three months. They scouted for new investors in the Los Angeles area, but found none. Everson had contacts at the Crocker Bank in San Francisco, so he set up a meeting for Farnsworth to make a presentation and then took him to a clothing store to buy him a respectable suit [7].

Philo Farnsworth

The Crocker bankers hired two engineers and they were impressed by the 19-year old who knew Einstein's theories and vacuum tube circuits. The group parted with $ 25,000 and offered space in a building where another Crocker venture was housed. They received 60% of the company and one of the bankers, Jesse McCargar, was named president. Philo Farnsworth was now a vice president and his ownership was down to 20% [12].

Farnsworth made the promise to the investors that he would be able to transmit an image within one year. The most important and most difficult part of his system was the image dissector tube; to do the glass blowing he hired his brother-in-law, Cliff Gardner, who had never done any glass blowing before but learned fast [20].

By the fall of 1927 Gardner managed to make a tube that transmitted a line; not much to look at, but an image nevertheless. By May 1928 Farnsworth had improved the apparatus to the point where he could show a still-picture (a triangle, to be exact), and then a photograph [7]. From there he moved to the projection of films. At first he had problems with sensitivity, the picture produced was too dark and too coarse; in a film of a hockey game, one couldn't see the hockey puck [5].

By the summer of 1928 Farnsworth had solved the problem of the sensitivity, gave a demonstration to the press and received considerable publicity. Using 150 lines, the picture was still poor (about 50,000 pixels) but faces were now recognizable.

But finances became critical. It had taken twice as long as promised and $ 60,000 (multiply these figures by a factor of about 10 to account for inflation between 1928 and 2006). The backers agreed to form a new corporation and raise additional capital. Their aim now was to sell the whole enterprise as soon as possible.

The relationship between Jesse McCargar, the president, and Philo Farnsworth, the vice-president became strained. Farnsworth would hire an employee, McCargar would fire him and Farnsworth would hire him right back.

Vladimir Zworykin

In April 1930 Vladimir Zworykin came for a visit. Zworykin was born in Russia and had studied under Boris Rosing, who had made a name for himself with his television experiments using a Nipkow disk. Zworykin emigrated to the United States and found a job at Westinghouse, working on an electronic TV system. He submitted a patent application in 1923, but then his project was canceled [1].

Zworykin was sailing under false colors. He had in fact just been hired by David Sarnoff to head a development program at RCA in competition with Farnsworth. The meeting was an uneven match: Farnsworth was 24, enthusiastic and open; Zworykin was 40 and held his cards close to his chest. The latter made only one slip: at the close of the meeting he said in his thick Russian accent, "I wish that I might have invented it" [21], a statement he was to regret soon.

Upon his return to the East Coast, Zworykin set up two groups at RCA, one to work on his own ideas, the other to duplicate Farnsworth's image dissector.

Times were bad. In October of 1929 the stock market had crashed and the Great Depression began. McCargar and the other bankers were pressing to sell the company; Farnsworth was hoping to license his patents and continue with the research. Finally, in June of 1931 they signed an agreement with Philco to fund the ongoing research and take out a license.

Philco was located in Philadelphia and Farnsworth, Pem and all five engineers moved there in July. Almost immediately there was a clash of cultures. Philco, being an East Coast establishment, insisted on formalities, the California crew did not. They were assigned space on the second floor, under the roof, no air conditioning, not even fans. The Farnsworth engineers stripped to the waist; the Director of Research came storming in and told them to put on shirts, ties and jackets at once. Cliff Gardner stood his ground and demanded a fan. They got their fan [7].

In May of 1932 RCA filed a patent interference case against Farnsworth. His 1927 patent application had been issued in 1930 while Zworykin's 1923 application still had not been allowed. RCA contended that the Farnsworth patent should be declared invalid because Zworykin's described the same system earlier.

In June Philco had an experimental TV transmitter. The company had tried to keep their television activity secret, but now the cat was out of the bag. RCA's television lab was located across the river, in Camden, New Jersey, where they had their own transmitter. The two teams could watch each other's pictures and get a good idea of progress made. Farnsworth's heart sank when he realized that the larger RCA team was rapidly gaining on him.

By 1933 Philco's support began to falter. They had been threatened by RCA with the loss of their radio-manufacturing license, if they didn't stop their TV project [12] and didn't get along well with the Californians. In September they parted company.

Farnsworth decided to stay in Philadelphia, where he could monitor RCA. He managed to get more money out of McCargar and improve his system. In 1934 he set up a TV studio at the Franklin institute and demonstrated it to the public for an entire week, with regular shows. His resolution was now up to 220 lines (about 200,000 pixels). Black and white of course.

In 1935 the patent office held a hearing on the RCA interference suit. RCA's main argument was that no boy of 15 would have been capable of conceiving such an advanced concept. Farnsworth's attorney found Justin Tolman, his old high-school teacher in Salt Lake City and brought him to Washington to testify. Not only did Tolman remember his unusually bright student, he had kept a drawing Farnsworth had given him [7].

When Zworykin took the witness stand, RCA's case began to collapse. He was vague and evasive and his lawyers didn't (or couldn't) present the apparatus he claimed to have built in 1923 or any notes he had made.

The patent examiners ruled in favor of Farnsworth. They even went so far as to state, "Zworykin's answers clearly suggest that his testimony is influenced by the present controversy" [21]. Translation: Zworykin had lied.

RCA appealed and lost again.

Farnsworth should have been overjoyed by the victory, but things were not going well. There were only two licensees, both in Europe, one in England, and the other in Germany. He went on an extended trip to Europe only to find that his equipment was lost in the fire of the Crystal Palace in London and the German company was ordered by the Nazi government not to pay the license fee. He fell ill and on his return he learned that McCargar had fired his entire crew. He moved to a farm in Maine with Pem to recuperate.

Farnsworth had one overriding problem, he couldn't relax. He was man driven by his ideas and he would never be completely well again.

Despite the decisive ruling by the patent examiners, Sarnoff was not ready to give up. In 1938 his lawyers found a Delaware court outside the patent system that would hear the case. This court ruled that Zworykin's application should be allowed, simply because Zworykin had invented *something*. Over the objection of the patent examiners the patent was issued, 15 years after the application had been submitted [21].

Farnsworth's financiers now had enough. The stock market was doing a little better now, money could be raised. They hired Edwin Nicholas, a former RCA manager, who in turn acquired an existing radio manufacturer in Indiana. Farnsworth supplied the name and was made vice president. He tried to help as much as he could, but his health was deteriorating. Nicholas, who knew RCA inside out, finally forced RCA to sign a licensing agreement; it was the first time RCA had ever paid royalties, so far it had only been collecting them.

On April 30, 1939 the World's Fair opened in New York and RCA used the event to launch television. Sarnoff was on the air, proclaiming that television was invented at RCA [21].

But it was a false start. On December 7, 1941 Japan attacked Pearl Harbor and the US was at war. All television transmission and development work ceased to make room for war production.

Farnsworth had a nervous breakdown and returned to Maine. He spent much of the war years there.

The new Farnsworth company did reasonably well with war contracts, but after the war it didn't have a peacetime product ready and fell into financial difficulties. Nearly bankrupt, it had to be sold to ITT and Farnsworth became an employee.

In 1947 Farnsworth's patent expired, just as TV was taking off. In 1946 some 5,000 TV receivers were sold in the United States. In 1947 production shot up to 160,000, in 1948 it reached almost a million, in 1949 three million and in 1950 seven million, half of the made by RCA.

On January 7, 1949 RCA's NBC aired a show entitled "Television's 25th Anniversary", celebrating Zworykin's December 1923 patent application. Sarnoff presented Zworykin as the "inventor of television" [21].

In 1950 Sarnoff lobbied the Radio Manufacturer's Association and got them to agree that henceforth he, Sarnoff, was to referred to as the Father of Television and Zworykin as the Inventor of Television. An RCA-wide memo ordered all employees to do so. Nobody dared to disagree.

At ITT Farnsworth moved into a new area. He convinced the management to let him work on nothing less than a novel nuclear power plant. His idea was to focus several of the same electron beams he had used in his TV work on one central spot, which, he was, convinced would cause nuclear fusion. His plan was to build millions of small reactors, just large enough to supply the power for a house.

To obtain fusion, the material at the center needed to be confined so that it didn't escape once it was heated up. Farnsworth believed he could do this with an electric field (i.e. electro-statically).

He spent several years on this project, but couldn't get it to work. He tried to get funding from the Atomic Energy Commission, but they declined. There were already several competing projects, using magnetic fields for confinement, or lasers to supply the starting energy. Known as Tokamak reactors (an acronym for "toroidal chamber in magnetic clouds", in *Russian*), some of them had achieved fusion. The development of the first Tokamak in the US cost $ 1.4 billion. Farnsworth, the lone inventor, never had a chance.

Farnsworth's health declined further. He had a perforated ulcer. Later he had a seizure and collapsed. He now weighed less than 100 pounds. In 1967 ITT stopped the project and put Farnsworth in medical retirement. He and Pem went back to Maine [7].

Incredibly he decided to start yet another company and continue the project two years later. The couple moved to Salt Lake City, mortgaged the house, rented lab space, hired a small crew and tried to raise money. But Farnsworth looked ill to potential investors and no money came in.

When they were unable to meet the payroll or make the lease payments and the payroll withholding taxes became delinquent, the Internal Revenue Service padlocked the lab and the bank loans were called.

In January of 1971 Farnsworth was hospitalized with pneumonia. His weight was down to less than 90 pounds [7]. He died on March 11, 1971.

Much Ado About *Almost* Nothing

10. A 30-Ton Brain

The dream to replace the human brain started almost 400 years ago.

And we are still dreaming.

The dismal, disastrous 30-year war was raging in Europe. Wilhelm Schickard, a professor for Biblical languages, astronomy and mathematics in Tübingen [1], had a revolutionary idea: he would build a machine that could add, subtract, multiply and divide. Some time before 1623 he commissioned a local mechanic to build his brainchild, and it worked. Then he had another one built for Johann Kepler. Unfortunately Kepler's machine was destroyed by fire before it was completed and the first machine fell victim to the war. It is only through Schickard's correspondence with Kepler that we know of Schickard's invention [11].

For addition and subtraction Schickard used a system of gears. Each wheel of the six digit machine had ten teeth, and the last tooth kicked the next higher wheel by one notch, thus "carrying" the ten [11]. The design of the calculator is remarkable not only because of its mechanical ingenuity, but for the radical departure from the commonly held contemporary belief that calculating could only be done by the human mind.

Schickard himself was killed by the plague in 1635 [1]. His work was not discovered until 1957, 333 years later and long past the time when his pioneering work could have been of use to anyone.

Blaise Pascal was born in 1623, the son of a tax court judge in Clermont-Ferrand, France. He was an odd child, slightly humpbacked and given to fits: his oddity was so marked that the family believed someone had cast a spell on him at birth [8].

Pascal was an extraordinarily gifted child who, at 16, made his first original contribution in geometry. At 19 he conceived of a mechanical calculating machine and began to develop it. It was not as advanced as Schickard's (of whom he hadn't heard), it could only add and subtract, but Pascal spent 10 years on his invention, building some 50 models and trying to produce it for commercial sale. By royal decree he was granted a patent (and thus a monopoly) on it [6].

Pascal's invention worked, but his design was taxing the mechanical skills of the time, which made the machine expensive to produce. Clerks and accountants worked for very low wages and against that background the mechanical calculator was a commercial failure [8].

Despite a painful and debilitating form of neural disease Pascal came up with more inventions (he conceived the wheelbarrow in its modern form and the hydraulic press) and produced first-class scientific work on the equilibrium of fluids and on the probability theory. At age 30 he renounced the vanities of the world, including science, and began to write on behalf of the Jansenists, who were involved in a bitter struggle with the Jesuits. Shortly before his death at age 39, Pascal started to work on a defense of Christianity. The work was never completed but his notes were published as "Pensees" [6].

Gottfried Wilhelm Leibniz was 23 years younger that Pascal. He had been born and raised in Leipzig, the son of a professor. He was one of those incredibly bright children, teaching himself to read Latin at eight and then going on to Greek. He received his doctorate at age twenty.

Being neither wealthy nor part of a religious order, he needed an income. The common thing for a man with Leibniz's intellect would have been to teach at a university. But to Leibniz such a career was too confining, so he attached himself to the nobility as a free-roaming thinker. And free-roaming Leibniz was; he seems to have been interested in everything: mathematics, politics, theology, mechanics, jurisprudence, linguistics, history, geology, metaphysics and philosophy. He conceived integral and differential calculus (as did Newton).

Leibniz had high hopes for mathematics, believing that its certainty would eventually eliminate the need for argument between philosophers. In 1672 he was sent on a mission to Paris, where he learned of Pascal's calculator. Enamored by the idea he improved the machine, adding multiplication and division to its capability. To do this he invented the stepped wheel, a cylinder with nine teeth of increasing lengths [21].

Again, however, there was no commercial success. Leibniz presented his machine to the Royal Society, and that was the end of the project.

Intriguingly, Leibniz later did pioneering work in binary algebra, the foundation of modern computers.

Charles Babbage, the son of a wealthy banker, was educated by private teachers and studied mathematics more or less on his own. When he entered Cambridge University he found that he knew more on the subject than some of his teachers [27].

Babbage graduated from Cambridge in 1814 at the age of 22. He was a gregarious, socially active young man with a delightful sense of humor. He got married, moved into a house in London and worked on whatever interested him, mostly mathematical problems [28].

Tables were an important part of mathematics in Babbage's time: multiplication tables, logarithmic tables, annuity tables, astronomical tables, and Babbage, with pedantic persistency, was forever finding errors in them. The more complex tables were calculated by division of labor: two or three highly competent mathematicians worked out the formulae to be used. A lower team of perhaps 10 mathematicians separated these formulae into routine calculations - mostly additions and subtractions - which then were assigned to a large number of semiskilled clerks. Often the labors of such teams took years [18].

It was this procedure that gave Babbage the idea that a machine could do the entire job of calculating tables, could do it faster and without errors. All he needed was a mathematical approach, which would let him break down the calculations into simple, routine calculations, which a machine could handle. This he found in the method of differences [13].

The method of differences (later known as the calculus of finite differences) uses an intriguing peculiarity of some functions, which is best explained with an example:

x	1	2	3	4	5	6
x squared	1	4	9	16	25	36
Difference 1		3	5	7	9	11
Difference 2			2	2	2	2

The third row is simply the difference between succeeding figures of the second row. The interesting row is the fourth one, where the differences (between the figures of the third row) become constant.

Now, working backwards, Babbage could arrive at the desired result (the second row) by simply adding the differences. He built a small prototype of such a machine, which had six decimal places. He demonstrated it to the Royal Society and then asked the government for a grant to build the real thing. His "difference engine" was to have no less than seven 20-digit registers and was to be capable of printing the results directly. He received his grant in 1823 and began to work. Highly skilled draftsmen and mechanics were hired. The project was to take two years - three at the most - but after four years only a small portion of the machine was finished. Babbage had greatly underestimated the complexity of the project and the precision required was at the edge of technology. In all, the drawings called for some two tons of precision gears made from brass, steel and pewter. But Babbage had a second problem: he was forever making improvements in the design, rendering much of what had already been done obsolete [28].

And now personal tragedy struck: his wife died giving birth to their eighth child [7]. Babbage had a nervous breakdown. He went on an extended trip and put the difference engine out of his mind. When he returned he pleaded for more money and the government obliged. By 1832 the government had spent £ 17,000, Babbage himself £ 20,000 and no difference engine was in sight [18].

In the midst of these cost overruns Babbage came up with a brand new idea, more grandiose than even the difference engine. He had learned of Jaquard's automatics loom, which employed punched cards to set the weaving pattern. Punched cards, he thought, could be used to tell a calculating machine not only what figures to calculate, but how it had to perform the calculation. This approach would make possible a machine that could calculate anything of any complexity. He dropped the difference engine completely and started to draw up plans for his "analytical engine". The project immediately grew to fantastic proportions. The analytical engine was to have a memory of 1,000 50-digit numbers, a central processor (which he called the mill), inputs through punched cards and outputs through printers. He planned to use a steam engine to drive it and, most remarkably, conceived of conditional branches, so that the machine could make its own decisions which way to proceed once it had come to intermediate results [29].

The government, now told that the difference engine was obsolete but that there was something much better, finally came to a decision: no more money for Babbage. It was a bitter disappointment. Gradually, Babbage became a cranky, bitter and frustrated old man, who felt that the world had done him wrong. He continued working on this and similar machines until his death in 1871, with money from his own pocket. But neither the difference engine nor the analytical engine was ever completed.

Ironically, a Swedish engineer, working from a magazine account, built a difference engine of more modest proportions - eight decimal numbers - during Babbage's lifetime; this machine worked and turned out to be quite useful.

Yet another mathematician: George Boole. He knew nothing about calculators or computers, but laid an important foundation for them. He was a simple, quiet man, who learned mathematics from his father, a cobbler. At age 15 he began to teach and by the time he was 20 he had his own small school. In 1849 he was appointed professor at Queen's College, Cork, Ireland, despite the fact that he had no formal education.

Boole wrote some 50 papers on mathematical research, as well as two textbooks and two volumes on logic. With these last two he made his most formidable contribution. Boole was able to write the laws of logic (known since Aristotle) in mathematical form. It turned out to be a very simple system: in Boolean algebra there are only two operations, "and" and "or", and there are only two outcomes, true or false (which turned out to be ideal for the binary digital computer).

But during Boole's lifetime his algebra remained a mere curiosity. Formal logics was not taken up again until some 60 years later [13].

In 1879 Herman Hollerith graduated from the Columbia School of Mines in New York and went to Washington, D.C., as part of a small army of people to work on the census of 1880. His special assignment was to prepare a report on steam and waterpower. It was a rather dull report that few people actually read. He stayed at this job for three years and then went to MIT as an instructor in mechanical engineering. But with him he took an idea. Census taking had become increasingly laborious and costly as the population of the United States increased. It was done by hand and there was a danger that it might soon take longer than the ten years between each census to get the results [2].

Hollerith's idea was to encode the information on each individual by punching holes into a paper-strip and then counting each attribute electrically with wires feeling the paper and making contact with a cup of mercury underneath when there was a hole present. In his spare time he started to construct such a machine. After a year he left MIT and became an assistant examiner in the U.S. patent office, a job he left less than a year later to become an independent patent consultant [2].

Bitten by the inventive bug, Hollerith suddenly started working on an electrically actuated airbrake system for trains. It was a disastrous project;

he couldn't compete with Westinghouse. But, in the meantime, he had seen a railroad ticket that contained some simple appearance information, so that the conductor could punch into the ticket a description of the owner, such as color of hair and eyes, which made the switching of tickets between several users more difficult. This ticket gave Hollerith the idea to use punched cards for his census machine, and back he went to the first project [2].

In 1886 Hollerith was able to try out his machine in the Baltimore census and soon small orders for machines to tabulate mortality statistics for the New York and New Jersey health department came in. But the big order Hollerith was hoping to get was for the 1890 census and he demonstrated, modified and improved his system until, in November 1889 he landed a contract to lease six machines to the Census Bureau for $1,000 per year each. This was followed by a contract for 50 more machines by January 1890 [2].

Hollerith's system consisted of two parts. A mechanical punch, operated by hand, was used to encode the card with the information, one card per individual. The second part was the counter. The card was placed into a jig and feeler wires lowered onto it. A large number of electromechanical counters were available to total up the counts, and the wiring of the machines could be changed to allow the freedom of which combinations were to be counted. Lastly, one of the feelers operated an electromechanical latch on one of 24 bins, into which the card was dropped by hand [13, 24].

Since 288 possible hole locations had been provided on each card, the Census Bureau was able to greatly expand the statistical work, while at the same time speeding up the count [2].

Hollerith then provided machines for Austria, Canada, Italy and for the very first census in Russia. In 1897 he incorporated the Tabulating Machine Company. The business grew steadily and profitably, supplemented by sales of cards in enormous numbers, cards that could be used only once. But Hollerith was the wrong man to manage a company. He couldn't delegate, was quick to anger and to take offense and had a tendency to alienate people [2]. Soon there was competition, notably by another former employee of the Census Bureau, James Powers, who had developed a machine that automatically printed results. The contract for the 1910 census was awarded to Powers [33].

In 1911 Hollerith sold out. Three companies were combined into a conglomerate, called the Computing-Recording-Tabulating Company. Hollerith received $ 1,210,500, built a large house on three acres in

Georgetown and bought a farm in Virginia [2]. Though his company was the smallest factor in the threesome, its business, properly managed, was to grow the fastest and become much better known by a different name: IBM.

In 1933 Konrad Zuse was a 23 year-old civil and mechanical engineering student at the Technische Universität in Berlin. Much of the work at school was pure drudgery, calculating stresses in structures using the same formulae over and over again. There were mechanical calculators available, in fact Germany was one of the most advanced producers of such machines. But it still meant sitting down for hours, going through the same routines. A much larger calculating machine was needed, he felt, which could be programmed to go through these routines automatically. He began to make sketches.

In 1935 he graduated and took a job with Henschel, an aircraft manufacturer. Calculating the stresses in aircraft wings turned out to be just as tedious and Zuse now became serious about his automatic calculator. After a year he quit his job to devote full time to the project. Reluctantly his parents set aside a room in their apartment as a laboratory and gave in to the idea that they would have to support their son for a while longer. Help, in the form of small sums of money and donated labor, also came from his former student friends.

Zuse's first approach was entirely mechanical (yet, he was unaware of Babbage's work). The program and data were entered through a hand-punched tape and the machine used the binary system throughout. The interior of the machine borrowed from developments in calculating machines and cash registers. The most remarkable part was the 1,000 bit mechanical memory, which required a volume of only 0.5 cubic meters (about 5 cubic feet), much less than an equivalent memory using electrical parts would have taken [5].

Gradually, however, Zuse the mechanical engineer began to see that electrical relays would operate better and faster, at least in the logic section, and he began to build a second version. Helmut Schreyer, a friend who spent his vacation helping, suggested using vacuum tubes. Zuse gave a paper at the university about his work in 1938, but when he mentioned the possibility that he and Schreyer might attempt building a machine with 2,000 vacuum tubes, there was general disbelief in the audience that such a thing could be done [41].

The political situation now started to interfere with the project. One of his helpers, who had joined the SS but then changed his mind, disappeared; Zuse was told that he had had an accident. In 1939 Zuse was called to active duty. He had with him a letter from the president of a leading calculator manufacturing firm, asking that Zuse should be excused from military service because he was working on a calculator for airplanes. The commanding officer looked at the letter and then exclaimed that such calculators weren't needed because German planes were already perfect.

After six months Zuse managed to get a release, going back to his old job at Henschel. He was assigned to do calculations for a remotely controlled plane. The development work on his second machine (which he called the Z2) continued evenings and on weekends and the machine began to work after a fashion, but it was quite unreliable because Zuse had to make do with second-hand telephone relays. Not only did he not have the money to buy new parts, there were simply no parts available without military priority [41].

Despite this shortage Zuse started to work on his third machine, the Z3, an ambitious project that would contain 2,000 relays. It was to be a computer with a floating decimal point, 22 bit words, a 64-word memory (still mechanical) and could multiply or divide in three seconds [26]. For the first time he received partial financing from the government, allowing him to start his own company.

In 1941 Zuse was called to active duty again, this time for the Russian front. Luckily, before he reached the front, Henschel managed to get him released. In addition to his government support for the Z3, he got an order from Henschel for a specialized calculator, one that would figure out the trim settings for a remotely controlled plane. This unmanned plane (or flying bomb), dropped and guided by radio from an aircraft, was to be mass-produced and Zuse's calculator used to determine the needed trim correction from measurements of the wings.

With income from the sale to Henschel Zuse started to work on his fourth machine in 1942 (the Z4), more advanced and larger than any of the previous ones. When the bombing of Berlin started, Zuse was forced to move his small staff and his machines three times, finally ending up in a basement. In the process the Z1, Z2 and Z3 were lost in fires.

Although Zuse was entirely apolitical during this time, one wonders, of course, what his motivation was. Long afterwards, when it had become fashionable in Germany to have been opposed to the Nazi regime, he recalled clearly that he wanted to contribute to the war effort to get back at the enemy who was bombing his German cities, especially after the bombing of Dresden, for which he couldn't see any military value. The more bombs fell, the more furiously he and his team worked [41].

Along came a second contract from Henschel, for a machine that could measure the deviations in the remotely controlled plane automatically and come up with the correction settings for the trim. This machine was to be installed in a Henschel plant in Czechoslovakia, but while Zuse was setting it up in early 1944 orders came through to dismantle the plant because the Allies were closing in on it. While the workers started to tear down the plant, Zuse worked day and night to get the machine operating. He got it working, went back to Berlin and never saw the machine again [41].

In the midst of all the bombing and confusion in Berlin Zuse decided to get married. As the Soviet army closed in, he managed to obtain transportation papers for himself, his wife, his workers and the Z4, but only because he had initially named his computer the V4 (for Versuchmodell 4 - research-model 4) and the official in charge thought that it was an advance V rocket. The group moved to Göttingen and the computer was in transit for two week, several times coming close to being destroyed. In Göttingen Zuse got the Z4 working for the first time [41].

As the Allies advanced toward Göttingen, Zuse found a way to attach himself to Werner von Braun's group, thereby gaining a truck and 1,000 liters of Diesel fuel. Traveling mostly nights over bombed out roads, they eventually reached Hinterstein in the foothills of the Alps near the Austrian border. The computer was hidden in a basement and remained undetected when French Moroccan soldiers occupied the area. Later two English officers, following a rumor that there was a secret weapon hidden in the basement, inspected it, but went away disappointed that it was only a calculator [41].

After the war Zuse's computer ended up at the Polytechnic University in Zurich. It turned out to be a highly reliable machine and, because the calculations took considerable time, it almost always ran day and night. It was the only computer then in existence in Europe.

George R. Stibitz was a "mathematical engineer" at the Bell Telephone Laboratories. In 1937 he was asked to look into the design of relays and he became enamored by the function of a relay as a logic element. Using a couple of relays from a junk pile, a scrap of board and some snips of metal he assembled a simple binary adder at home on his kitchen table [26]. His colleagues at Bell Labs were amused that relays could be used in this way but didn't see any significance at first.

Stibitz, however, realized that this element could be enlarged and become very useful. Stibitz convinced the management that a relay calculator would save considerable time. He designed such a machine in 1938 and Western Electric built it. When completed in early 1939 it performed the calculation three times faster than a row of women with mechanical calculators [37].

Stibitz was a shy, retiring man. He requested an allocation of $ 20,000 for a bigger and more automatic machine, but was turned down [26]. Assigned to work for the Army during the war, he designed five more specialized relay computers, but this work led him away from the main path of computer development [5].

In contrast to Stibitz, Howard H. Aiken was definitely an outspoken man. He was an associate professor of mathematics at Harvard in 1936 when he began thinking about building a computer. Aiken had studied Babbage's work carefully and it seemed to him that calculator technology had advanced to the point where it should be possible to build a machine similar to the analytic engine. But Harvard was an unlikely place to find support for the building of a giant calculator; it was a center of pure research and the opposition was strong. Some held that, if Aiken got such a machine working, it would only be a short time before it would have performed all the work it could possibly be required to do [36].

Aiken wrote a proposal and took it to the Monroe Calculating Company, whose management decided the project was too impractical and turned it down. Then he went to IBM, where he found a sympathetic ear. Aided only by a minor financial support from the Navy, IBM undertook to build a computer for Aiken at no charge to Harvard University. Although the "Automatic Sequence Controlled Calculator" was Aiken's brainchild, IBM engineers did much of the design [37].

Construction of the machine started in 1939 and took nearly five years. It was a machine to behold. Driven by a five horsepower motor, a shaft extended through almost the entire 17-meter (50-foot) length of the shiny, stainless-steel clad 5-ton monster, turning counter wheels with electromagnetic clutches through chains and sprockets. The computer was capable of 23 decimal digits and had 72 adding storage registers, each with 24 counter wheels (23 for the digits and one for the sign). Controlled by perforated tape, it could do an addition in about three seconds. The entire machine contained almost a million individual components and sounded like a room full of ladies knitting [5, 32].

Unfortunately Howard Aiken was not a very diplomatic fellow. He renamed "his" computer the Harvard Mark I and at the dedication ceremony scarcely mentioned IBM, which made the representatives from IBM furious. The Mark I received an inordinate amount of publicity, and for the first time the term electronic brain was mentioned.

Electronic it wasn't. For all the hoopla the Mark I operated very slowly and was to be outdated in less than two years by a computer that would surpass it in speed by a factor of 1,000.

John Atanasoff was a professor of physics and mathematics at Iowa State College at Ames. He had started to think about building a computer as early as 1935. His idea was to use vacuum tubes, which could operate at a much higher speed than either mechanical counters or relays. By 1937 he had settled on the binary approach (Aiken, like Babbage, used the decimal system) and a year later he had most of the details worked out. He asked for a grant of $ 650 from the college to build a small prototype, which he received early in 1939. With this money he was able to hire an assistant and the two of them had the prototype working in the fall. It was not a full-fledged computer, but it proved that vacuum tubes could indeed be used for the purpose and performed at the speed Atanasoff had predicted [29].

In designing his vacuum tube computer, Atanasoff made an invention that turned out to be a fundamental one for computers: the dynamic memory. He decided to use capacitors to store the data - one capacitor for each bit. But capacitors do not hold a charge indefinitely; there is always some leakage current which tends to gradually reduce the voltage. Atanasoff's solution to this problem was elegantly simple: he cycled the data through a chain of capacitors, each time refreshing the charge. In this way the capacitor only needed to hold its charge for one cycle [12].

In 1940 Atanasoff wrote an unpublished paper about his idea and used it to raise money to build a real computer. He asked for $ 5,000, but only managed to scrape up $ 1,460. Nevertheless he and his assistant started to build a large machine, one that could solve 30 simultaneous differential equations. In 1942, with only parts of the computer operational, Atanasoff went to work for the Naval Ordnance Lab and the machine was abandoned. Iowa State never bothered to apply for any patents [12, 37].

During the summer of 1941 a young professor of physics named John Mauchly went to visit Atanasoff and saw what was possible to achieve with vacuum tubes. Mauchly had received a scholarship for Johns Hopkins University and went directly for a Ph.D., without taking the exams for either a bachelor's or a master's degree. When he finished his studies in 1932, jobs were extremely scarce. For a year he worked as research assistant (at 50 cents an hour), a job, which required the constant, and tiring use of mechanical calculators. In 1933 he became a professor of physics at Ursinus College near Philadelphia and got interested in statistical predictions of weather. He attempted to churn the massive data collected from weather observations with his mechanical calculator (a task which even overwhelms a modern computer even today). So when Mauchly went to see Atanasoff he was eager to learn about the speed of vacuum tubes [8].

During the same summer Mauchly also took a course in electronics at the Moore School of Electrical Engineering of the University of Pennsylvania. His laboratory instructor was J. Presper Eckert. Eckert, who had received his B.S. that year, was a gifted and skilful engineer. The two of them started talking about the possibility of building vacuum tube computers. Mauchly received an invitation to join the faculty of the University of Pennsylvania and joined in the fall [37].

Again, Mauchly's new job involved mechanical calculators, lots of them. The University of Pennsylvania had a contract from the U.S. Army Ordnance Corps to calculate the trajectories of shells. These calculations needed to be done for every type of gun and shell and it took hours to come up with a single set of data. The procedure was for a mathematician, such as Mauchly, to break the task down into small steps of routine calculator work, which was performed by dozens of female clerks. Mauchly, convinced that the entire task could be done in a fraction of the time by a single, large machine, wrote a proposal with Eckert in 1942; neither of them knew of Babbage, Zuse, Aiken or Stibitz. The Army turned the proposal down, believing that the war was going to be over before the machine could be put into operation [37].

By 1943, however, the situation had changed. It no longer looked like the war would end soon and the calculating backlog had increased intolerably. A young lieutenant and mathematician, Herman Goldstine, assigned by the Army to the Moore School, asked Mauchly and Eckert to resubmit their proposal [36].

What Mauchly and Eckert proposed was audacious. Their first estimate was for a machine with 5,000 tubes, costing $ 140,000. (The most complex electronic apparatus built up to that time had 400 vacuum tubes). But the Army wanted more computing power and the project grew to a computer with 18,000 vacuum tubes, which would eventually cost $ 400,000. Most experts ridiculed the project, expecting the limited life and general unreliability of the vacuum tube to make the computer in-operational most of the time. Enrico Fermi, who had had unpleasant experiences with some of his own vacuum tube equipment, estimated that such a complex computer would not work for more than two minutes before one of the vacuum tubes failed [8].

The Army, however, went ahead; a contract was signed in April 1943. A team was assembled at the Moore School, but despite a furious pace the ENIAC (for Electronic Numerical Integrator and Computer) did not get completed until a few months after the war ended. Nevertheless, it was a substantial achievement.

Filling an entire, large room and weighing 30 tons, the ENIAC performed 5,000 additions or subtractions per second and consumed 150 kilowatts of power [5, 32]. Compare this with your average laptop computer today: 5 pounds, 50 million additions or subtractions per second and 50 watts.

Most importantly, the ENIAC was remarkably reliable, working about 90 percent of the time. The reliability was almost entirely due to Presper Eckert's careful work. Not satisfied with ordinary design methods, he had analyzed each component and designed the computer so that it would still work when the components were at the end of their life. When the machine was moved to Aberdeen Proving Grounds, however, the ENIAC suddenly became very unreliable, working less than 50 percent of the time. This sudden decline in reliability was puzzling, until Eckert and Mauchly found out that Army regulations required the machine to be shut down at night, unless a guard was present. In the morning it took several hours to replace the vacuum tubes, which had failed when power was turned on again [26].

The ENIAC, for all its speed, had one serious, glaring shortcoming: to do a specific task, a panel needed to be rewired, a job, which could take several hours (it also used the decimal rather than the binary system). During the building of the ENIAC a new idea came up, an idea, which provided a significant advance in computer design: the stored program. In this concept the program is entered into the computer as if it were data, stored in the memory of the computer and called up by the computer whenever it is needed. It is still not entirely clear who invented the stored program concept. Eckert and Mauchly claimed the invention for themselves, but so did another man: John von Neumann.

Herman Goldstine met von Neumann at the train station of the Aberdeen Proving Grounds in 1944 and, both having security clearances, got to talking about ENIAC. Von Neumann, who was born in Hungary, was an exceptionally brilliant man and had an incredible memory, making it appear that he remembered everything. Goldstine once asked him if he knew the beginning of The Tale of Two Cities. Von Neumann not only knew the beginning, he continued reciting the book until Goldstine begged him to stop after 15 minutes [16].

When Goldstine met von Neumann, the latter was a professor at Princeton, a consultant to the atom bomb project and had a need for computing power. He wrangled a visit to the Moore School and attached himself to the team. Sometime during the uneasy collaboration between Mauchly, Eckert and von Neumann the idea of the stored program was born and a new computer, the EDVAC (Electronic Discrete Variable Computer) was started. Von Neumann, together with two coauthors, published the idea first and without giving due credit, which led to bitterness between him and Eckert and Mauchly.

The stored program concept turned out to be of key importance. The controversy of its authorship is further complicated by the fact that the idea had already been implied in a 1936 paper by the brilliant and eccentric Alan Turing (who, at roughly the same time as Eckert and Mauchly, pioneered vacuum tube computers within the British war effort). Ten years before the EDVAC, Turing had already come to the conclusion that data and instructions are fundamentally identical and that, by having both readily accessible in memory, the power of a computer could be increased by a very large factor [5, 36].

Ironically the EDVAC was not the first stored program computer in operation. This honor belongs to Maurice Wilkes and his EDSAC (Electronic ..... never mind!). Wilkes had worked at the Moore School in 1946 and went back to the University of Manchester with his own ideas for a stored-program computer. He was able to beat Mauchly, Eckert and von Neumann to the punch because the Moore School team was rapidly disintegrating [5, 8].

Apart from the clash of personalities, the disintegration was brought on by a desire of Mauchly to reap monetary rewards. He had been spending a considerable portion of his time on Wall Street trying to raise money for a company. The university was unhappy about this situation, but there was a more serious problem: Mauchly and Eckert felt the patents resulting from their computer work belonged to them. The university disagreed; in March 1946, one month after the dedication of the ENIAC, the university asked all employees to sign a patent release for future work. Eckert and Mauchly refused and were forced to resign. Eckert got immediate job offers from von Neumann and IBM, but Mauchly convinced him to help him start a company. Mauchly had already started negotiations with the Census Bureau for the development of a large, more advanced computer [34].

But money proved to be hard to raise and the negotiations with the government were tedious. The Census Bureau called in the National Bureau of Standards for advice, which in turn formed a committee of experts, including von Neumann, George Stibitz and Howard Aiken. It was not a good start for the new company. Von Neumann was clearly hostile toward Eckert and Mauchly. Stibitz called the project naive and too ambitious. Aiken's opinion was that two of these proposed computers would suffice to do all the computing ever needed, thus using much of the same argument that had been used against him nine years before [34].

Nevertheless a contract was signed by the Census Bureau, but it was only for a study phase, covering expenditures of $ 75,000. It took a year to complete this phase, during which time Eckert and Mauchly were operating at a loss. Then the report had to be submitted to the committee again, which meant a further delay. In the hope of raising venture capital the Eckert-Mauchly Computer Company was incorporated, but no venture capital was found. Strapped for cash, Eckert and Mauchly accepted a contract from Northrop to develop a small, special purpose computer for the Snark missile, which gave them an $ 80,000 down payment on a $ 100,000 project. In June 1948 a $ 169,000 agreement with the Census Bureau was finally signed. In addition, two commercial sales of the "Univac" computer were made and, for a brief time, the future looked bright [34].

But all of the contracts Eckert and Mauchly had signed were for fixed prices, for machines, which had not been developed yet. And, as you have probably already suspected, Eckert and Mauchly grossly underestimated costs. In the nick of time the American Totalisator Company, a maker of racetrack equipment, agreed to finance the company in the form of $ 500,000 in loans, for which it received 40 percent of the stock of Eckert-Mauchly Computer Co. The loans patched up the company and it looked like Eckert-Mauchly might be a success after all. But, in October 1949, the vice president of American Totalisator in charge of the deal was killed in a plane crash and American Totalisator called the loans as of January

1950. Eckert and Mauchly scrambled to either raise money or find a buyer for the company. The only offer they received was from Remington Rand which, in February 1950, absorbed the Eckert-Mauchly Computer Company. Eckert, Mauchly and some of the employees received a total of $ 100,000. In addition Eckert and Mauchly were to receive 59 percent of the net profits obtained from their patents [34].

With their patents the two entrepreneurs were unlucky too. In 1973, as a result of a dispute between Sperry Rand and Honeywell, the main patents were declared invalid because they had been anticipated by Atanasoff [34].

John Mauchly later tried to get into business for himself one more time and failed again. Eckert stayed an employee and became fairly wealthy.

We now return to the Computing-Recording-Tabulating Company, the conglomerate that gobbled up Herman Hollerith's company. Its sales amounted to less than $ 4 million and the prospects were somewhat less than exciting, until Thomas Watson arrived.

Watson had joined the National Cash Register Company as a salesman in Buffalo, New York, when he was 21. NCR was the creation of John Patterson, an erratic but highly creative tycoon. Patterson ran NCR with an iron hand, indoctrinating rather than training his salesmen, requiring his employees to be well dressed all the time and never to imbibe in public. He also posted slogans everywhere and fired his top staff periodically.

Watson, a star salesman, rose through the ranks and ended up at company headquarters in Dayton, Ohio. In 1903 he was put in charge of a scheme to eliminate some of the competition. A second-hand business for cash registers had sprung up, small outfits that undersold NCR with refurbished models. It was Watson's job to set up stores, undersell the used-model dealers, put them out of business and then pull out. All this was to be done in a way that made it appear Watson's operation had no connection with NCR [33].

Watson did his job well. So well, in fact, that a competitor sued NCR under the Sherman Antitrust Act and the federal government joined in the suit. Early in 1913 Watson and Patterson were both convicted and sentenced to a $ 5,000 fine and a year in jail. Both appealed. Toward the end of that year Patterson had one of his frequent tantrums and house-cleanings and Watson, at age 40, was fired; as he left NCR he vowed to build a business bigger than Patterson's [3, 33].

So, when Watson joined the Computing-Recording-Tabulating Company, he faced the prospect of spending a year in jail. But an appeals court ordered a new trial and the government dropped the suit. Now Watson went to work, building a sales force in the image of NCR, updating the products (especially the long-neglected Hollerith machines) and posting slogans everywhere (including the well-known THINK, which had already been in use at NCR). Like Patterson, Watson was strong-willed and had a flaming temper, but he was far more successful at keeping valuable employees. By 1917 sales had increased to over $ 8 million and in 1924 the company was renamed International Business Machines [33].

By the beginning of the Depression, sales of IBM had more than doubled again, but then stayed flat at the $ 20 million level until 1935. It was in 1939 that Watson made good on his vow and overtook NCR in sales. By this time Hollerith's former company was a dominant division within IBM, mostly renting card punches, sorters (which could handle up to 400 cards per minute), and accounting machines, which could produce a printout from the cards. In addition IBM had started to manufacture electric typewriters [33].

IBM's main competitor was Remington Rand. Neither company pushed into computers after the war; in fact, there was very little commercial interest in computers among manufacturers. It was only through Watson's interest in educational institutions that IBM supported Aiken's efforts and then, in revenge for Aiken's slighting of IBM's role in the Mark I development, came up with a faster electromechanical computer in 1948, but built only one [37].

Watson, now 74, had lost his zest for change; it was time for a successor. The job fell to son Thomas Watson jr. and, when, it was he who took steps for IBM to enter the field of vacuum-tube computers. IBM's first entry was unremarkable. Remington Rand, who had acquired the Eckert-Mauchly Computer Company, had a head start. Their Univac computer could be used for both business and science, while IBM's computer was strictly a scientific machine [32, 33]. For the next five years the Univac remained the best computer on the market [9].

But Remington Rand stumbled badly. Indecision and corporate intrigue rendered it immobile; it merged with Sperry and was renamed Sperry Rand. IBM on the other hand was well managed had a well trained and disciplined sales force. By 1980 IBM had almost two-thirds of the computer market, while Sperry shared the rest with six other competitors [9].

The age of the vacuum tube computer lasted less than ten years. Transistors provided higher reliability and smaller size and then integrated circuits became the driving force, shrinking the computer's size decisively, while reducing its power consumption and increasing its speed by large factors. Thus the story of the computer becomes the story of semiconductors, which we shall explore in the next chapter.

Where does it all lead? With the appearance of the electronic computer, science fiction writers found a seemingly inexhaustible subject: a computer more powerful than the human brain. Can the computer replace the brain? Will there eventually be a computer more intelligent than a human being?

Developing such a computer would be a task more formidable than any science fiction writer ever imagined. The human brain contains some 100 billion neurons [19] (a rough estimate; no-one has actually ever been able to count them). Each neuron is a highly complex structure, consisting of a body, the axon, and hundreds or thousands of fibrous branches, the dendrites. The cell wall of the neuron acts as a pump, separating sodium and potassium ions and thus acting like a battery with a very small voltage.

The dendrites and the axon are connected to other neurons through synapses, giving a total of 100 trillion connections [17] (again a rough estimate). This incredibly complex structure weighs 1.5 kilograms and consumes 25 watts [31].

When the dendrites receive enough signals from neighboring neurons, the small potential inside the neuron collapses and the change travels long the axon, delivering a signal to other neurons. The speed of this transmission is surprisingly slow, about 30m/sec (90 feet/sec) and takes place in a rather odd form: the nerves send pulses at rates varying from 2 to 100 per second, with the rate indicating the intensity of the experience. The main signal paths in the brain are, therefore electrical but, at the synapse, communication becomes purely chemical (or, more accurately, electro-chemical): across this narrow cleft molecules of many different kinds are transmitted [37].

Even though the computer is purely an electrical machine and the brain employs a mixture of electricity and chemistry, the basic elements of the two are the same. They are both gates, responding to stimuli. But there the similarity ends. Computers use simple gates (with an average of perhaps three inputs), which behave according to strict logic rules. In the brain this gate is far more complex (up to 10,000 inputs) and it has a logic form far more flexible (some inputs say yes, some no and some perhaps or almost) [19]. This is the main reason the brain is less accurate than a computer, but far more creative. Also, in a computer the logic wiring is fixed, in the brain it is subject to learning. Each time we experience something, new connections are formed; repetition strengthens these connections and transforms then into long-term memory [19]

But the greatest difference between brain and computer is the way in which information is stored. The brain does not use ones and zeros in a hierarchy of files and directories. When you try to recall the name of an old acquaintance there is no alphabetical entry. Instead there is an impression, an image, consisting perhaps of a visual picture, sound or even smell [15]. This is the main reason why 40 years of work on artificial intelligence have shown almost no progress; the computer simply does not function like the human brain.

Ray Kurzweil has written an intriguing book [22] in which he predicts that, with the rapid and enormous increase in the complexity of integrated circuit the computer will soon match the capability of the human brain. The question is, will the computer then be the same as a human brain and *replace* it?

And why would you want the replace the human brain with a computer, when the former can be duplicated by relatively unskilled labor?

Much Ado About *Almost* Nothing

11. Noyce

The genius who became a failure,
eight smart men,
and an inventor created by lawyers.

All his life Ferdinand Braun was scholarly and diligent. At age 15 he wrote an entire textbook on crystallography with some 200 drawings in his own hand, but publishers shied away because the author was so young. He received a Ph.D. magna cum laude at age 22 [11].

In **1874**, when he was 24, Braun made a discovery. He attempted to measure the electrical properties of some naturally grown crystals and had a great deal of trouble making contact to them. He finally got the setup working by winding a wire around the bulk of the crystal and shaping a second wire like a coil spring, letting the tip of the wire press against the surface of the crystal [11].

What he measured was very confusing: the material did not obey Ohms law. Some 50 years earlier Georg Ohm had found that an electric current was always proportional to the voltage. Like water, if you increased the voltage (the height of a reservoir), the current increased proportionally. The two were related by the resistance; in an electrical conductor the resistance is determined by the thickness, length and material of the conductor, in a river by the width, depth, length and any obstruction in it.

What Braun measured was quite different. Not only was the current not proportional to the voltage but, when he reversed the battery, he got a different result. He published his findings, but neither he not anyone else could explain the effect or find any use for it (the deviation was only about 30%). It would lay dormant for 30 years [11]. Little did Ferdinand Braun know that he had stumbled onto the world of semiconductors, which would eventually explode into a giant industry.

Later Braun invented the cathode ray tube (CRT) as we saw in chapter 9 and then went on to investigate wireless transmission, designing apparatus for transmitting underwater, which was not a success. But in 1898 he worked out a novel design for a spark transmitter, using tuned circuits, which was considerably better than any transmitter Marconi had been using.

Braun was doing this work as a consultant to an investor group. A company called Telebraun was formed as the equipment designed by Braun began to exceed the performance of the Marconi stations. Siemens became a major investor and soon Telebraun was, for all practical purposes a subsidiary of Siemens.

There was competition within Germany: AEG had also become interested in wireless and had acquired the patents of other inventors. The Kaiser was mightily displeased by the competition of two major companies within his own country and ordered the two efforts to be merged, forming Telefunken. As his name disappeared from the company logo, Braun became relatively obscure, though his technical contributions exceeded those of Marconi and were considerably more fundamental. In 1909 Braun shared the Nobel Prize for physics with Marconi [11].

Braun's life ended in sad circumstances. In 1914 he went to the United States to defend one of his wireless patents in a lawsuit with the American Marconi Company. He waited too long to return and when the United States entered the war he could no longer find passage home. In 1917 his wife died back in Germany and in April 1918 he himself succumbed, in a foreign country at war with Germany [11].

Nothing much happened until about 1904, when radio appeared on the scene and a "detector" was needed. The signal was "amplitude modulated", which simply means the strength of the signal fluctuated in rhythm with the speech or music. Our ears cannot hear radio frequencies, so a device is required which removes the radio frequency and leaves only the modulation. The simplest way to do this is to cut the radio signal in half (i.e. "rectify" it) and then smooth the waveform. Braun's device did exactly that; it let one half of the signal pass, but blocked the other half.

Thus, 30 years after Braun's discovery, the "odd behavior" of a wire touching a crystal (such as Galena, silicon carbide, tellurium or silicon) found a practical application [15]. The device was called the "Cat's whisker". It did the job, but not very well; one had to try several spots on the crystal until one was found which produced a loud enough signal.

And it was replaced almost immediately by the vacuum tube, which could not only rectify but amplify as well. Thus the semiconductor rectifier (or diode) went out of fashion.

About 15 years later the rectifier reappeared in a different shape: bulky, messy contraptions using copper-oxide (and later selenium) to produce DC from the line voltage, chiefly to charge car batteries. But nobody had any idea how or why these things worked.

Then World-War II happened and with it came radar. To get adequate resolution, radar needed to operate at high frequencies. This time a world-wide emergency drove the effort, with plenty of funding for several teams. In the U.S. the government gave contracts to MIT, Purdue University, The University of Pennsylvania, GE and Bell Labs. They started with the "cat's whisker" and tried to determine what made it so fickle and unreliable.

It became immediately obvious that the materials used contained all sorts other substances. They found a way to get rid of the contamination: they heated a small portion of the crystal close to the melting point and then moved this zone through the crystal in one direction, repeatedly. The impurities stayed in the hot zone and thus were eventually all collected at one end of the crystal, which could be cut off and discarded. They became so good at it, produced such pure material, that Braun's effect disappeared. Now they realized that some of these impurities were actually *required* to make a diode. And these impurities all fell into very specific places within the periodic table of elements.

Then they remembered that an A.H. Wilson had written a paper 15 years earlier, which explained how, semiconductors worked, a paper that had been largely, ignored [15].

To keep the explanation simple we concentrate on the two elements, which soon became the favorites. **Silicon** and **germanium** both have a valence of four. Valence simply means that, in the outermost layer of electron orbits, there are four electrons. Silicon, for example, is element number 14, meaning it has a total of 14 electrons. The first orbit (or energy level) has two electrons, the second eight and the third four.

The outermost orbits of the atoms touch each other and the electrons in this orbit don't stay with one particular atom, they move around the neighborhood. As an electron circles its atom it comes to a point where it is equally inviting to continue the circle or move around the neighboring atom. Thus in a crystal, where all atoms are perfectly aligned, electrons are not loyal to one atom, but constantly switch from one atom to another. It is this sharing of electrons that hold the atoms together. And this ability to move from atom to atom is also the basis of electrical conduction: in conductors the electrons roam widely and are easily enticed to move in an electrical field, whereas in an insulator they stay close to home.

Electrically, pure silicon is a terribly uninteresting material. It is an insulator, but not a very good one. The fun begins when we add the right impurities, or dopants.

Just to the right of silicon in the periodic table is **phosphorus**, element number 15. Like silicon, it has two electrons in the first orbit, eight in the second but there are five in the third. Now let's say we were able to pluck out an atom in a block of silicon and replace it with a phosphorus atom. Four of the valence electrons of this new atom will circulate with the silicon electrons, but the fifth one won't fit in. This excess electron creates a negative charge and the silicon becomes what we now call n-type.

This introduction of excess electrons is unlike static charge. When you brush your hair so that it stands upright, you have simply moved some electrons temporarily. When you "dope" silicon, the charge is permanent, fixed in the crystal lattice (but does *not* become a battery).

Similarly, to the left of silicon and one space up in the periodic table is **boron**, element number 5. It has two electrons in a first level and three in a second, a valence of three. If we replace a silicon atom with a boron one, there is an electron missing and we create a positive charge, or p-type material.

Even when a semiconductor is doped n-type or p-type, it is still an uninteresting material. It is neither a good conductor nor good insulator. But when you place n-type and p-type regions close together, something very unusual happens.

Imagine you are a wizard who can replace some silicon atoms at will with either boron or phosphorous atoms. You made a silicon crystal with a sharp demarcation line in the middle; to the left the silicon is n-type, to the right p-type. Within the first micro-second or so some of the excess electrons on the left will move across the demarcation line and fill the voids of the missing electrons (the holes) on the right. But they will do this only for a very short distance (a few micro-meters), as far as their electric fields reach. Then everything stops. The demarcation line is now in a no-man's land, a neutral region where there is neither an excess or a deficiency of electrons (called the depletion region).

Now let's apply a voltage across the silicon. If he n-type region is made positive and the p-type one negative, nothing much happens. The potential simply pushes the electrons and holes in the silicon farther apart and the depletion region widens. No (or very little) current flows.

If, however, the p-type region is made positive and the n-type one negative, the electrons and holes at the boundary of the depletion region are pushed closer together and some of the electrons and holes (collectively called "carriers") jump across the gap and current can flow.

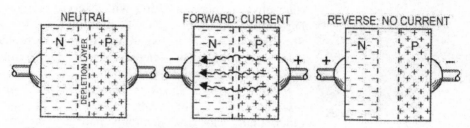

The Diode (or Rectifier). When P and N-doped regions are adjacent, some electrons fill holes, forming a no-man's land (the depletion region, left). With a voltage applied in the forward direction, the depletion region narrows until it is small enough that a current can flow (center). If the voltage is reversed, the depletion region widens and no (or very little) current flows (right). Thus a diode conducts in one direction only.

You have to keep in mind that very few dopants are required to create this effect, perhaps one dopant atom for every 100,000 silicon atoms. Naturally grown crystals contain foreign material in much greater numbers, which is the reason the effect was so difficult to observe or duplicate.

With this understanding the wartime teams were able to make diodes that were far more predictable and reliable than the cat's whisker. And, while the cat's whisker and the copper-oxide or selenium rectifiers were based on an odd effect between a semiconductor and metal, these diodes were using a pure element (germanium at first, silicon later).

The story now focuses on Bell Labs. And it is here that we meet the driving force for the next stage, William Shockley. His father was a mining engineer, his mother a mineral surveyor. Their jobs required them to pull up their roots frequently and Bill was born in London in 1910, the parents' only child. When Bill was three years old, the family moved to Palo Alto, California, where he was educated at home until he was nine.

Bill was obviously a bright child; he would frequently get into deep conversations with a neighbor, a Stanford physics professor. But the parents moved again, to Southern California. Bill went to Hollywood high school and then on to the California Institute of Technology, where he studied physics.

In 1932, with a B.SC., he went to MIT on a teaching fellowship and emerged four years later with a Ph.D. Then he joined Bell Laboratories, which had already become one of the choicest places to do research. Shockley moved around within Bell Labs, getting involved in many areas, including vacuum tube technology and solid-state physics, but a meeting with Mervin Kelly, the Director of Research, stuck in his mind. Kelly told Shockley that what he wanted most was to replace switching contacts in telephone exchanges with a more reliable electronic switch [18].

William Shockley

Around 1938 Shockley became involved in a research project with Walter Brattain. Eight years older than Shockley, Brattain had grown up in Washington State on a farm. After earning his Ph.D. he spent one year with the National Bureau of Standards, then joined Bell Labs in 1929 as a research physicist and began to investigate semiconductors [4].

It was the depletion region that intrigued Shockley when he started to work with Brattain. It occurred to Shockley that, if he could somehow insert a grid into this region, it might be possible to control the amount of current flowing in a diode, creating the solid-state equivalent of the vacuum triode, the very device that was on Kelly's wish list [18].

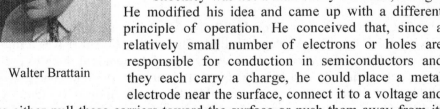

Walter Brattain

Shockley went to Brattain with the idea and Brattain was amused. The same idea had occurred to him too; he had even calculated the dimensions for such a grid, which turned out to be impractically small. Shockley tried it anyway and couldn't make it work. Brattain had been right [5].

Shockley was not a man easily defeated, though. He modified his idea and came up with a different principle of operation. He conceived that, since a relatively small number of electrons or holes are responsible for conduction in semiconductors and they each carry a charge, he could place a metal electrode near the surface, connect it to a voltage and thus either pull these carriers toward the surface or push them away from it. Therefore, he thought, the conduction of the region nearest the surface could be altered at will. He tried it -- and it didn't work either.

Unknown to Shockley at least three other people had come up with the same idea before him, all with disappointing results. Between 1925 and 1928 Julius E. Lilienfeld, a native of Poland and professor at the University of Leipzig, applied for three patents on the same structure. It is not clear whether he ever got his devices working. In one of the patent applications he shows a diagram using this device in a radio; if the device worked at all it clearly would have been far too slow for such a use [3].

The quest for a semiconductor triode stopped abruptly at the outbreak of World War II. Both Shockley and Brattain went to work on military projects.

After the war Shockley and Brattain returned to Bell Labs. Shockley was made co-supervisor of a solid-state physics group, which included Brattain. The progress made in refining silicon and germanium was not lost on Shockley; he decided to try his idea for an amplifying device again and had a thin film of silicon deposited, topped with an insulated control electrode. It still didn't work; no matter what voltage was applied to the control electrode, there was no discernable change in current through the silicon film. Shockley was puzzled; according to his calculations there should have been a large change. But the effect - if there was any - was at least 1500 times smaller than theoretically predicted [18].

It was at this time, in 1945, that John Bardeen joined Shockley's group.

Bardeen became the theorist of the group. He looked at Shockley's failed experiment and mulled it over in his head for a few months. In March 1946 he came up with an explanation: it was the surface of the silicon, which killed the effect. Where the silicon stops, the outermost four electrons can no longer share their orbits with neighboring atoms; they are left dangling. Bardeen correctly perceived that this incomplete structure at the surface produced charge (or voltage), which rendered any voltage applied to an external control electrode ineffective. It would take many years to overcome this surface effect.

John Bardeen

With this theoretical breakthrough the group now decided to change directions; instead of attempting to make a device, they investigated the fundamentals of semiconductor surfaces. It was a long, painstaking investigation; it took more than a year. On November 17, **1947** Robert B. Gibney, another member of the group and a physical chemist, suggested using an electrolyte to counteract the surface charge. On November 20 he and Brattain wrote a patent disclosure for an amplifying device as tried by

Shockley but using electrolyte on the surface. Then they went to the lab and made one. The electrolyte - glycol borate, called "gu" - was extracted from an electrolytic capacitor with a hammer and nail. The device worked on December 4, the electrolyte did precisely the job that Gibney thought it would [18].

But, although this "field effect" device amplified, it was very slow, amplifying nothing faster that about 8Hz. Brattain and Bardeen suspected that it was the electrolyte that slowed down the device so, on December 16., they tried a different approach: a gold spot with a small hole in the center was evaporated onto germanium, on top of the insulating oxide. The idea was to place a sharp point contact in the center without touching the gold ring, so that the point would make contact with the germanium, while the insulated gold ring would shield the surface. But in the process of trying this a voltage discharge spoiled the hole. They placed the point contact very close to the periphery of the ring and suddenly got amplification [4].

There was only one thing wrong with this device: it didn't work as expected. A positive voltage at the control terminal increased the current through the device when, according to their theory, it should have decreased it. Bardeen and Brattain investigated and found they had inadvertently washed off the oxide before evaporating the gold, so that the gold was in contact with the germanium. What they were observing was an entirely different effect, an injection of carriers by the point contact. They realized that, to make such a device efficient, the distance between the two contacts at the surface needed to be very small, in the order of 0.05 millimeters. They evaporated a new gold spot, split it in half with a razorblade and placed two point contacts on top. Now the device worked even better and they demonstrated it to the Bell management on December 23. 1947 [4, 18].

A few days later John Pierce, Executive Director of the Communication Sciences Division, sitting in Brattain's office, named the device. He felt the name should be similar to varistor and thermistor, two then promising devices, so he suggested calling it the transistor [4].

For half a year Bell kept the breakthrough a secret. Bardeen and Brattain published a paper on June 25. 1948 [1] and on June 30 a press conference was held in New York. The announcement made little impression. The New York Times, for example, buried the news in brief item on the foot of the radio section on page 46 [5].

Shockley had been disappointed by the turn of events; he had not been part of the final breakthrough. But he realized that, even though there was a working device, the battle wasn't over yet. No one within the group really understood precisely how the transistor worked. So, in the early days of

January 1948 Shockley sat down and tried to figure out what was going on between the two point contacts. He devised an experiment to gain more information and in the process, on January 23, suddenly conceived of a much better structure: the junction transistor [18].

It is now time to take a look at how a transistor works and we shall use Shockley's early junction device as a model (the point-contact transistor operates the same way, only the geometries are different).

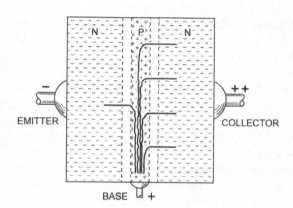

The Junction Transistor. There are three layers, with the center one (the base) doped oppositely from the outer ones. The base-emitter junction is a forward-biased diode and a current flows. But, because the base is very thin, most of the carriers (electrons or holes) are swept away to the collector and never reach the emitter. Thus a small current controls a much larger one and the device has a "gain".

Let's start with the same diode, an n-type region on the left, p-type on the right. Now let's make the p-type region very thin and add another n-type region on he right. We now have an NPN sandwich with two demarcation lines and two depletion regions close together. Now let's make three contacts, two at opposite sides to the n-regions, one (a bit more tricky) at the bottom to the thin p-region. We call the left-hand region the emitter, the thin center the base and the right hand region the collector. Then we apply voltages: Both the base and the collector are positive; the emitter gets the negative terminal. Current thus flows in the original diode. As the carriers flow through the base region they have to pass by the depletion region created by the addition of the collector. If they get too close to this depletion region they are swept away by the collector voltage and end up flowing through the collector terminal rather than the base terminal. The trick, then, is to make the base region very thin so that the majority of the carriers flowing through the base end up in the collector. It is as if these carriers were promised to travel to the base terminal but on the way are hijacked by the collector. In a modern transistor as many as a thousand of these carriers are sucked up by the collector for every one that does manage to reach the base terminal, thus giving the device a "current gain" (or amplification) of 1000. By varying the current between the base and emitter terminals, a much larger variation of the collector-emitter current is created. It is also possible to reverse all the polarities, ending up with a PNP transistor.

But it was easier to design such a junction transistor on paper than to make it work in an actual piece of semiconductor. Shockley wrote several papers and an entire book on its theory before another Bell Labs researcher, Gordon Teal, found a way to make a practical device possible.

Gordon Teal had joined Bell Labs in 1930. Both his master's and doctoral theses had been on germanium. At Bell Labs he first worked in the chemical research department, then in the electro-optics department. World War II, however, brought a de-emphasis of television work and Teal was transferred back to the chemical research department to work on germanium and silicon for radar diodes. It was Teal who made the thin film of silicon for Shockley's first postwar experiment [19].

Teal was dissatisfied with the state of the art of removing undesired impurities from semiconductors. At the time of the invention of the transistor the annual world production of germanium was only about 6 kilograms (12 pounds). Teal wanted to do more research and he wrote memorandum after memorandum, proposing new methods to refine these materials. More importantly, Teal was firmly convinced that, to obtain optimum performance, semiconductors needed to be single-crystal. The point-contact transistor had been made from polycrystalline germanium, i.e. it had a structure containing many small crystals without a common orientation [19].

No one paid much attention to Teal. Shockley remained unconvinced that single-crystal material was necessary and all of Teals memos were properly filed and forgotten.

But, on September 30 1948, around quitting time, Teal ran into John Little, also with Bell Labs. They chatted and it turned out that Little needed a germanium rod of small diameter. Teal said that he knew how to make such a rod and that it would have the unique property of being a single-crystal. On the bus to Summit, New Jersey, they continued talking and sketching an apparatus. Little happened to have a bell jar with a radio-frequency heater. They decided to melt the germanium in a graphite crucible and use a clock mechanism to pull the rod out of the melt. A small piece of crystal was to be dipped into the molten material and, when pulled out slowly, the germanium would solidify in single-crystal form (the basic method of crystal pulling had been invented in 1918 by a man with the tongue-twisting name of Czochralski). Without bothering to ask for permission Teal and Little set up the apparatus and two days later were able to pull the first single-crystal germanium rod [19].

As Teal had suspected, this material indeed improved the performance of the point-contact transistor, yet there was still skepticism. After two months funds were finally allocated to make it an official project, but no laboratory space was made available. Teal and Little had to set up their crystal puller at night in a large metallurgical shop and then roll it into a closet when the day shift arrived [19].

Teal now suggested to Shockley that a junction transistor could be made by pouring small amounts of impurities into the melt while the crystal was being pulled. Shockley, having no alternative, started to collaborate with Teal. The melt was first made p-type. After the crystal had been pulled part way, n-type "dopant" (i.e. an element with a valence of five) was added in sufficient quantity to outweigh the effect of the first impurity. Then, very quickly to make the base region thin, the melt was converted back to p-type by pouring more dopant into the melt, thus producing a PNP structure in the crystal rod. The rod was then sawn into small bars and wires were attached to each of the three regions [19]. By April 1950 the first junction transistor was working [18].

Just in time. The point-contact transistor had been giving headaches when Bell started to produce it. It was maddeningly difficult to accurately place the points and make sure they wouldn't move. The junction transistor, although slower, was far more predictable.

But even so, transistor production grew slowly. As a replacement for the vacuum tube the transistor was a disappointment for the first few years. Not only was it slower and considerably more expensive, it was also different. The vacuum tube was, conceptually, a voltage amplifier: a small (voltage) signal at the grid resulted in a large (voltage) signal at the plate (or anode). The transistor, on the other hand, was basically a current amplifier and any attempt to treat it like a vacuum tube was bound to fail. Thus, a generation of electronic engineers who, by now, had gotten used to the vacuum tube as the central element in electronics were asked to change career-long habits.

There was also the question of reliability. One of the main attractions of the transistor was the absence of a filament, which took time to heat up and would eventually burn out. The transistor had no such limitations, it could last forever - or so it was thought until production started and the annoying fact came to the surface that even a "solid-state" device could be as unreliable as a vacuum tube. To be sure, these failures were not due to an inherent burnout mechanism, but they were failures nevertheless, created by mechanical weaknesses, metallurgical imperfections or chemical contamination.

There were two additional problems. Despite utmost care it was difficult to control the performance of the transistor. Small variations in the material or the processing temperature had magnified effects on the transistor's gain, for example, requiring the devices to be separated into several classes after manufacture, some of which were of little use. Worse, a significant percentage of the transistors didn't work at all and there was little chance of repairing them. Thus a term came into use, which has stayed with the industry, ever since: yield. By the time the transistor can be tested it is too late to do anything about defective parts, they have to be discarded. And, sometimes, for unknown reasons, this yield would dip as low as a few percent.

But, in spite of its teething problems, the transistor had powerful advantages which could not stay ignored for long: it was small, required much less power, didn't need to operate at high temperature and had no warm-up time. AT&T decided to license the manufacture of transistors to anyone who could come up with a $ 25,000 advance payment on royalties and, in 1952, held a seminar to teach its licensees how to manufacture them. In memory of Alexander Graham Bell, the teacher of the deaf, transistors for hearing aids could be made royalty-free. (In 1956, however, AT&T was forced to sign a consent decree, surrendering all existing transistor patents to the public) [5].

Most licensees went to established vacuum-tube manufacturers - General Electric, Westinghouse, Raytheon, Philco, Tung-Sol - who saw the new device as a possible threat to their business. Even before the license had been offered, the news of the Bell Labs development had triggered research efforts in many of these companies. In 1950 General Electric came up with a different way to make a junction transistor. Starting with a thin wafer of germanium, indium (p-type) dots were alloyed from both sides, leaving a narrow n-type (base) region in the center. Philco, in 1953, improved on GE's technique by thinning the wafer with a jet of acid before alloying, so that the thickness of the n-region after alloying could be controlled to an extreme degree [5,7].

In 1952 Gordon Teal left Bell Labs and joined Texas Instruments. This company (between 1930 and 1950 known as Geophysical Services and engaged in oil exploration) had decided to branch out and signed up for a transistor license. Teal knew exactly what he wanted to do: make a silicon transistor. Compared to germanium, silicon remains a semi-conductor over a wider temperature range and is thus better suited for military and demanding industrial applications (germanium transistor circuits would often quit working when exposed to the hot sun). But to get the same performance, the base in a silicon transistor needs to be even thinner than that of a germanium device [19].

Teal knew how to grow crystals and he chose the same technique he had successfully used to make the first junction transistor for Shockley. Because of an inherent difference between the two materials, it was easier to make an NPN transistor in silicon (vs. a PNP transistor in germanium) and Teal found a way to make the doping time for the middle region very brief, resulting in a thin base region [19].

In 1954 the first silicon transistor was ready and TI decided to announce the breakthrough at a meeting of the Institute of Radio Engineers (IRE) in Dayton, Ohio. It turned out to be one of those situations every inventor or researcher dreams of, but few have the pleasure of experiencing. Teal's paper was the last one of the morning session and several of the preceding speakers expounded in great detail on the difficulties of making a silicon transistor and stated that it would be at least several years before one would see such a device. When Teal stepped on the podium he pulled a few of his silicon transistors out of his pocket and nonchalantly described how he had done it [19].

When, in 1956, the three inventors of the transistor were awarded the Nobel Prize for physics, only Walter Brattain was still at Bell Laboratories (he retired in 1967 and became a visiting professor at his alma mater, the Whitman College at Walla Walla, Washington). John Bardeen had left in 1951 to become a professor at the University of Illinois and, for his research there in superconductivity (extreme conductivity at very low temperatures) he received a second Nobel Prize in 1972 [8].

Bill Shockley left Bell Labs in 1954; it was not an amicable parting [15]. There were complaints about Shockley's prickly personality and Bardeen and Brattain no longer wanted to work for him. In a reorganization Shockley lost most of his group.

He spent one year teaching at Cal Tech and worked for the Pentagon for another. Then found a backer in Arnold Beckman of the Beckman Instruments Company. He divorced his wife and moved to Palo Alto, California. A subsidiary, called the Shockley Semiconductor Laboratories, was set up.

His task now was to staff the new company. He called former acquaintances for recommendations and ran ads. Applicants had to submit to psychological testing, have their IQ determined and solve mathematical puzzles devised by Shockley. Within a year he had some 20 people - predominantly Ph.D.s, and none from Bell Labs - working for him, among them:

• Victor Grinich, who had received his Ph.D. from Stanford University and had worked as a research engineer at Stanford Research Institute. Grinich found an advertisement of Shockley's in a professional magazine, written in code. The code intrigued him so much that he had to find a way to decipher it and then applied [9].

• Robert Noyce, who had met Shockley at an industry conference while employed by Philco. Noyce, born in 1928 and raised in a number of small Iowa towns (his father was a Congregational preacher), developed a love for mechanical gadgets. While still in high school, Noyce took physics courses at Grinnell College. As it happened, the physics professor there had gone to school with John Bardeen and he was able to obtain one of the very first transistors. It was this early sample of a transistor which caused Noyce to select semiconductors as his field of specialization. After college he went to MIT, received a Ph.D. in physical electronics in 1953 and then joined Philco [22].

• Gordon Moore, the son of an under-sheriff in San Mateo County. After a Ph.D. in chemistry and physics Moore worked at the Applied Physics Laboratory of Johns Hopkins University, from where he joined Shockley.

• Jean (translated: John) Hoerni, born 1924 in Geneva, Switzerland, who had received not just one but two Ph.D.s, one from the University of Geneva, the other from Cambridge. He had worked three years for Linus Pauling at Caltech doing X-ray diffraction work, but gradually had become dissatisfied when he realized that the chemists made all the glamorous investigations, while he was mired in routine analysis work. Hoerni applied at Bell Labs for a position with Shockley, but found that the latter had already left. When Hoerni sent his application to Palo Alto he was quickly hired, even though he had never worked on semiconductors [9].

For all of these people there was a brief period of fascination after they joined. But then the true Bill Shockley appeared from behind the glitter of fame and they discovered that Shockley was, in fact, a rather erratic and unpleasant man. He was a micro-manager who would fire his employees for minor mistakes, throw tantrums over trivial problems and change directions for no apparent reasons. He incessantly tried innovative management techniques, such as posting everybody's salaries on the bulletin board [9].

Hoerni, for one, found himself isolated in a former motel room across the street, doing theoretical calculations. When he complained about it he was summoned, together with two other Ph.D.s who had also incurred. the wrath of Shockley. The three of them were given an assignment: develop a four-layer diode - a silicon transistor with an additional layer, but only two terminals - within six weeks, precisely according to a Bell recipe, without any help from technicians or other personnel. They did it, but the episode was one more step toward a brewing mutiny. Not only was Shockley a very difficult man to work for, but his crew began to feel that events were bypassing them [9]. The four-layer diode, for example, didn't seem to have a future.

At Bell Laboratories, in the meantime, an entirely new way to get the desired small quantities of impurity atoms into the semiconductor had been developed: diffusion. At room temperature gases mix even if there is no turbulence. This happens because each atom or molecule moves around randomly due to the energy it receives by temperature. The higher the temperature, the more pronounced is this movement and thus the mixing. If the temperature is high enough (e.g. over 1,000°C) such gases can even diffuse into solid material, though their diffusion speed decreases enormously. Thus, for example, a p-type semiconductor exposed in a high-temperature furnace to n-type impurity atoms (in gas form) develops an n-layer at its surface with a depth as far as the impurities penetrate (and exceed the p-type impurities already in place). The main advantage of diffusion over growing or alloying is the accuracy with which a layer can be made.

Noyce and Moore were pushing Shockley to make silicon transistors using this diffusion approach, which promised improved speed and gain. Shockley wasn't interested; his hope was for his laboratory to come up with an entirely new device, a device that would represent as large a step over the transistor as the transistor had been over the vacuum tube [9].

Now totally dissatisfied, the crew talked to Arnold Beckman, the president of the parent company, and informed him of the impossible situation. Beckman promised to hire a business-minded individual who could

Robert Noyce

act as buffer between Shockley and his staff. But the solution didn't work; Shockley refused to let go of the day-to-day decision-making. Out of patience, seven staff members -Victor Grinich, Gordon Moore, Jean Hoerni, Jay Last, Sheldon Roberts, Julius Blank and Eugene Kleiner - decided to either join a different employer as a

Gordon Moore

group or start a new company. Eugene Kleiner knew someone at Hayden, Stone & Co., the investment broker, who began to look around for start-up capital. At the last minute Noyce, who was still hopeful to reconcile things with Shockley, joined the group. A deal was reached with the Fairchild Camera and Instrument Company, which had a desire to diversify into the transistor business. In October **1957**, the group departed [9].

The new company, called Fairchild Semiconductor, was at first an independent operation, with Fairchild Camera and Instrument holding an option for a buy-out. None of the founders had any management experience (six were scientists, two engineers); they decided to place themselves under a general manager and hired Ed Baldwin. The product they began to develop was the one they had proposed to Shockley. The detailed structure of this device, called the Mesa transistor, had been tried in germanium before, but not in silicon. It required two diffusions, both into the same side of a silicon wafer. The first diffusion was p-type, the second n-type, and the difference in depth between the two layers created the base region that, for the first time, could be made with a high degree of accuracy. The top surface of the transistor was then masked with wax and the exposed silicon etched away, giving the remaining piece a mesa-like shape. (A mesa is a flat-topped hill seen in the American South-West).

Because of its superior performance, sales of the Mesa transistor took off almost immediately. In October of 1958 Fairchild Instrument and Camera exercised the option; Fairchild Semiconductor became a wholly-owned subsidiary and each of the founders received $ 300,000 in Fairchild stock [9].

But there were also problems. The most serious one concerned the reliability of the Mesa transistor. The etched silicon chip was soldered onto the bottom of a small metal case, leads were attached to the top regions and then the case was welded shut. Tiny metal particles, ejected during the welding process, floated around inside the case and kept on shorting out the exposed P-N junctions [9].

Jean Hoerni was in charge of the physics section of the research and development department at Fairchild. Although he was aware of the reliability problem, his main project was to produce a PNP transistor with the same methods.

Silicon immediately grows a thin oxide layer when it is exposed to air. This is simply glass (silicon-dioxide) and its growth can be enhanced by moisture at high temperature. A few of the dopant gases used in diffusion (such as gallium) can penetrate this oxide layer, while most are stopped by

Jean Hoerni

it. There is, therefore, a possibility that the oxide layer can be used as a mask. If the oxide was to be etched off in some places but not in others and suitable dopant gases used, diffusion would take place only in the areas without oxide. A study done at Bell Labs had shown that phosphorus, for example, would diffuse only part way into the oxide layer. But the same study also came to the conclusion that an oxide layer exposed to a diffusion is left contaminated and must subsequently be replaced by a freshly grown one [9].

This bothered Hoerni. He didn't see any reason why the oxide layer could not be used as a diffusion mask for both diffusions and why the oxide should subsequently be regarded as contaminated. So he tried it - as an unofficial side project - and out of the trial came an advance ranking in importance second only to the transistor itself: the "planar" process.

In preparation for the first diffusion Hoerni spread a photosensitive and etch-resistant coating (called photoresist) over the top of the oxide and exposed it through a photographic plate (a mask) carrying the patterns of the base regions; such photographic techniques to mask and etch patterns had already been employed to fabricate "printed" circuit boards. The subsequent etching then removed the oxide only in the regions where p-type impurities were to be diffused. After the first diffusion he closed these oxide "windows" again by placing the

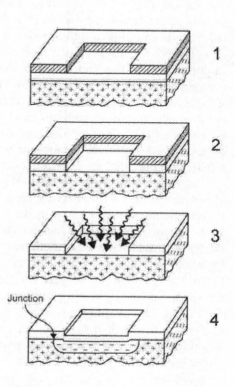

Hoerni's planar process, one of the most important inventions of the 20th century. In step 1 a photosensitive layer is exposed to light through a mask. In step 2 the oxide layer covering the silicon wafer is etched away in the exposed area. After the photosensitive layer is removed, impurities are diffused through the opening (step 3), thus forming a protected junction underneath the oxide. In step 4 the opening is closed by growing more oxide.

wafer in high-temperature moisture and then repeated the steps for the second (emitter) diffusion. In a third masking step windows could then be etched in the (re-grown) oxide to make contact to the two diffused layers (a contact which was most reliably established by evaporating aluminum onto the top surface of the wafer and then patterning the aluminum with the same photographic techniques as used before). The wafer could then be scribed (like glass) and broken into individual transistor chips.

The planar process had a whole series of advantages. Of most immediate importance was the fact that the oxide, one of the best insulators known, protects the junction. No longer could the metal particles from the welding of the case short it out. The second advantage was in the way the regions were created: photographic methods could be used to delineate not just one but hundreds of transistors at the same time on a silicon wafer. Thus individual, delicate masking of each transistor was no longer required, giving the planar transistor a potential for greatly reduced cost.

It was at this point that Ed Baldwin, the general manager, decided to strike out on his own. After barely a year with the group he left and founded Rheem Semiconductor to compete with Fairchild. It was the first of many spin-offs to come, but the hapless Baldwin made the wrong decision: he felt he wanted to start with something proven and geared up to produce Mesa transistors. The planar transistor left Rheem at the starting gate; it soon had to be sold to Raytheon [9]. Noyce, who took over from Baldwin at Fairchild, saw the advantage of the planar process and quietly moved it into production.

There was another advantage to the planar transistor: once the dopant enters the silicon it diffuses in all directions, including sideways. The P-N junction, therefore, ends up underneath the oxide and is never exposed to either human handling or the contamination of air. For this reason the planar junction became the cleanest (and most stable) junction ever produced [9].

Five years after the transistor had been invented, G.A.W. Dummer of the Royal Radar Establishment at Malvern, England, gave a speech in Washington, D.C. Dummer had a dream: the transistor produced its effect in a solid block of material, so why not try to implement all electronic functions in solid blocks, without any interconnecting wires [5].

Dummer didn't propose any concrete ways to achieve this, but after another five years of Dummer's missionary zeal and prompted by the shock of Sputnik, the Royal Radar Establishment gave a contract to Plessey to develop this concept, while the U.S. Air Force entered into a contract with Westinghouse to achieve the same goal. The U.S. Signal Corps, also feeling pressure to decrease the weight and size of its equipment, gave a contract to

RCA. Employing only proven fabrication methods, the approach to be developed under this contract re-packaged electronic components in "micro-modules".

A great deal of money was spent by the military, but neither approach led to the next generation of electronics. The molecular electronics program was far too ambitious, calling for a command performance of innovation, which never came. The micro-module program, on the other hand, was too timid. Although these programs undoubtedly provided some of the impetus, the solution to making electronic equipment smaller, lighter and more reliable was found outside any military program.

Jack Kilby had graduated in electrical engineering from the University of Illinois in 1947 and had joined Centralab in Milwaukee, Wisconsin. In 1952 Centralab acquired a license from Bell Labs to manufacture transistors and Kilby, a husky-voiced, large-framed Kansan, became involved in semiconductor technology. In May 1958 he joined Texas Instruments and was assigned rather vaguely to the area of "microminiaturization". In July the plant shut down for annual vacations and Kilby, who, as a new employee, had no vacations coming, was left alone to ponder over what approach to take. It occurred to him that portions of a transistor also exhibited the properties of two other electronic devices, resistors and capacitors. A piece of semiconductor, for example, has a certain amount of resistance, determined by the amount of doping. Though this resistance cannot be controlled very accurately, it is resistance nevertheless. The junction of a semiconductor diode or the metal-oxide structure on the surface of a transistor has an (undesired) capacitance, albeit a small one.

Kilby conceived that these two additional effects could be useful and that, together with transistors and diodes they could make a reasonable variety of circuits, circuits in which all the components consisted of one material only, either germanium or silicon [10].

Since all of the components were to be made from the same block of material, Kilby needed a way to isolate them from each other and then connect their terminals together. And for this he chose a less than inspired way: he separated them by either sawing or etching deep grooves and then used gold wire to make the connections. It was a production nightmare and the drawings in his patent application reminds one of Rube Goldberg (a cartoonist famous at that time for his drawings of absurd machines).

While Kilby was working on his circuits in Texas, a similar but far more elegant idea occurred to Robert Noyce in California. Noyce's motivation was primarily cost, not size. He realized that it didn't make sense to fabricate transistors precisely arranged in a pattern on the wafer, then cut them apart, place them in a housing and arrange them again in some order on a circuit board. If the additional components on the circuit board could be placed on the wafer, a considerable number of manufacturing steps could be saved. Noyce had no problem visualizing capacitors and resistors made in silicon, he was constantly dealing with these (unwanted) effects. What was needed, though, was an inexpensive way to interconnect these various components on the wafer. The idea of using wires had no chance in Noyce's mind; it would have simply been too expensive. But when Jean Hoerni explained his planar process to Noyce, the latter saw the solution: the aluminum layer used to make contact to the base and the emitter of the transistor could also be used to interconnect different components [9].

On January 23. **1959** Noyce entered his idea into his notebook. There was nothing he could do immediately; the planar process didn't exist yet. But when Texas Instruments announced Kilby's "solid circuits" at the IRE Show in New York, Noyce hurried to write a patent application.

Kilby had filed a patent application on February 6. 1959, Noyce on July 30. Kilby's patent attorney had urged him at the last minute to add language saying that the components could also be interconnected in other ways, such as laying down a gold layer. Kilby didn't say how this layer, supposedly adhering to the complex, three-dimensional surface could be patterned to form interconnection stripes.

Because of this additional language the two applications were clearly in interference. In 1962 a bitter battle between the two companies started in the courts. The first round went to Texas Instruments. But, on appeal, the decision was overturned and priority was granted to Fairchild and Noyce [10]. Texas Instruments appealed, but the Supreme Court refused to hear the case.

While the case was stuck in court neither company could collect royalties. In 1966, before the final verdict was in, the managers and lawyers of Texas Instruments and Fairchild Semiconductor got together and made a deal: Kilby and Noyce were to be declared co-inventors of the integrated circuit and the royalties split.

Thus we have the case of the two inventors of the integrated circuit: Kilby, whose complicated approach never had a chance of succeeding, and Noyce, whose elegant solution is present in every single integrated circuit even now.

But the real hero of the story is Jean Hoerni, who is virtually unknown but whose planar process has become the basis for the entire semiconductor industry.

Ironically, Jack Kilby won the Nobel Prize for his "invention" in 2000, after both Bob Noyce and Jean Hoerni had died.

By rewarding the founders of Fairchild Semiconductor early, Fairchild Camera and Instrument inadvertently did itself a disservice. The founders were now employees, in a business environment, which made it attractive to start new companies. Not surprisingly, an exodus of founders followed. In 1961 Jean Hoerni and Jay Last left to form Amelco, a division of Teledyne. Though the venture was a disaster and Hoerni left after only two years, he kept on starting new companies and amassed a tidy fortune [9]. He died in 1997. By 1968 Fairchild had 15,000 employees, but only two of the eight founders were left, Noyce and Moore and they decided that it was time to start a new venture: Intel.

The original idea for a semiconductor amplifying device still looked attractive to many people: make two small n-type regions on a p-type chip, spaced a short distance apart. Then place some metal on top, insulated by the ever-present oxide layer. A positive voltage on the metal (the gate) will attract electrons to the surface, which converts the area underneath the gate to n-type, i.e. the two regions are now connected by an n-type channel. Such an MOS (metal-oxide-semiconductor) device promised to be considerably smaller than the rather complicated bipolar transistor. And, like the NPN/PNP transistors, it seemed a small thing to invert all polarities and make complementary (CMOS) devices (with a p-channel transistor on the same chip), which would automatically lead to greater speed and lower power consumption.

But moving the concept into practice turned out to be incredibly difficult; it took 25 years after Shockley first tried to make such a structure for MOS to compete with bipolar transistors and resulted in disaster.

The first company at the starting gate was General Micro Electronics (GME). Receiving funding from Pyle National, a connector manufacturer, the company was founded by Arthur Lowell (a retired Marine Corps Colonel and former head of the Avionics Section of the Navy Bureau of Weapons), Howard Bobb (who had been in transistor marketing at Hughes), Robert Norman (a circuits and systems designer at Fairchild) and Phil Ferguson (who had been making transistors at Texas Instrument and then Fairchild). They set up a plant in Mountain View, in Fairchild's vicinity, and as their first product they elected to manufacture a special kind of (bipolar) integrated circuit, a derivation of Fairchild's first micrologic circuits, modified for lower power consumption. Both Lowell and Bobb were convinced that there was a need for such a product at the National Security Agency, which, they projected, would buy these circuits in large quantities and at high prices. But the National Security Agency, being involved in secret matters, could not divulge its purchasing plans and, being a government agency, could not move with any discernable speed. By 1964, when it became clear that sales for this product might be a long time in coming, GME needed some other product badly. It was then that Phil Ferguson proposed developing MOS integrated circuits [9].

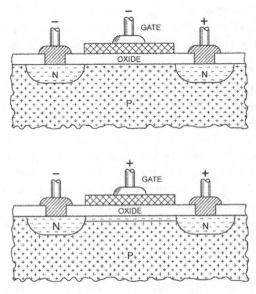

The MOS Transistor. It was the first device that Shockley (and others) tried, but it took an incredible 25 years to make it work. The idea is very simple: place two doped regions some distance apart in a piece of silicon which is oppositely doped. No current can flow between the regions since at least one junction is reverse-biased. Then apply a voltage to piece of metal (the gate) which attracts electrons to the surface, converting a thin strip to n-type and thus providing a channel for the current. Shown is an N-channel transistor, but P-channel devices are also used. Instead of metal, silicon is now used for the gate.

GME now got a development contract from Victor Comptometer (a comptometer was a large, mechanical calculator used in banks) and the initial results were impressive. While, in 1964, a micrologic-type integrated circuit contained some 20 to 30 components, the simpler process and smaller component dimensions allowed GME to place up to 400 components on an MOS chip. Thus the ten circuit boards of such a calculator could be reduced to one, with a corresponding reduction in equipment size and cost [9].

But disaster struck almost immediately. In producing these MOS circuits GME discovered an effect that was to plague the industry repeatedly and cause a number of production lines to be shut down, including GME's. Some unwanted elements can find their way into the oxide layer, the worst of which is sodium. Even in minute quantities, sodium can gravely affect the performance of an MOS transistor. Worse, it is relatively mobile, which means it can drift in and out of the oxide layer, changing the performance of the devices from hour to hour or day to day. Thus, when the device or circuit is tested, there is no assurance that it will work reliably or work at all.

The solution to the problem, painfully developed over the next few years, was extreme cleanliness in the manufacturing operation. But GME didn't survive to benefit from either the solution or the enormous later success of MOS integrated circuits. Lowell departed toward the end of 1964 and the company was sold to Philco in January 1966, having only made chips for some 2000 calculators. Phil Ferguson became a consultant; Howard Bobb and Robert Norman both founded their own companies, both making MOS integrated circuits. Bobb's American Microsystems succeeded, Norman's Nortec failed.

What happened to Shockley? His company, renamed Shockley Transistor Corporation, limped along for another two years after the departure of the eight Fairchild founders, without announcing any breakthroughs or showing a profit. Then Beckman Instruments sold it to Clevite Transistor for $ 1 million. Clevite, losing money itself, sold it to ITT in 1965, which closed it down in 1969 [5].

Shockley left before the final shutdown and became a professor at Stanford University. His main interest now shifted from semiconductors to sociology and he developed a theory called "dysgenics". What Shockley found was that people with lower IQ's are poor, poor people have large families; therefore, this segment of the population breeds itself to a lower and lower level. Because of the large contingent of blacks among poor people, Shockley's theory raised controversy; its validity hinged mostly on the accuracy of measuring IQ without any racial bias and the arrogant and abrasive Shockley was the wrong man to make a convincing case.

He came to believe that dysgenics was his greatest contribution to mankind, but friends and foes alike saw it otherwise; when he died in 1989 his reputation was utterly destroyed.

By the time Robert Noyce and Gordon Moore left Fairchild to form Intel, the basic trend in semiconductors had become clear. Steady improvements in the photographic techniques allowed the dimensions of the devices in integrated circuits to be made ever smaller, while gradually improving processing know-how made it possible to fabricate larger and larger circuits without defects. The first integrated circuits contained no more than 10 components (and were difficult to produce); by 1968 integrated circuits with a few hundred components were in routine production.

Noyce and Moore felt that this evolution was bringing semiconductor technology toward opening a huge market in computer memories. At the time of the formation of Intel the central memory of a computer was made from large arrays of small magnetic cores (i.e. rings), through which were woven (by hand) criss-crossing wires. Current through the wires magnetized or de-magnetized selected cores, thus leaving them in one state or the other. The cost of such memories was roughly one cent per core (i.e. per bit of information). Intel's first semiconductor memories, which contained up to 1024 bits, were still somewhat more expensive than cores, but had the advantage of being much smaller in size. By the time the complexity could be increased one more step, however, the integrated circuit memory was a clear-cut winner and cores quickly disappeared from the market [9].

Shortly after the start of Intel, Ted Hoff joined the company as employee number 12. Hoff had received both an M.S. and a Ph.D. degree in electrical engineering from Stanford University and had stayed on for six years as a research associate. At Intel his specialty was applications engineering for memories - helping customers understand the novel products [9].

Just as Intel got into business, electronic calculators exploded onto the scene. The point had been reached where electronic calculators could be made more cheaply than mechanical ones - three years after GME's heroic but ill-fated entry. A Japanese company asked Intel to fabricate the integrated

Ted Hoff

circuits for an ambitious new calculator it had under development and Ted Hoff got involved in the negotiations with the customer. Hoff didn't like the design of the calculator; it required 12 chips and it seemed to him that there should be a less complicated way to do it. He had been working with computers already back at Stanford and was familiar with their architecture. Drawing from his experience, he proposed a radically different approach in which one fairly large chip did all the computing, simulating the heart of a larger computer (the central processing unit, or CPU). The function of this chip could be altered by command from the outside. The Japanese engineers disagreed with Hoff.

Hoff went to Noyce, who told him to go ahead and design such a processor. In October 1969, going over the heads of the engineers, Intel made a presentation to the management of the Japanese company and received a contract for Hoff's design [9].

Forming a team with two other engineers, Hoff now converted his design to integrated circuit form (with 2,300 components), a pain-staking, year-long effort [9].

But Hoff wasn't satisfied making this circuit for only one user. His familiarity with computers told him that such a processor should be useful in many applications and the team started pestering Intel's marketing department to get a release from the customer and sell the chip as a product. The marketing department was not interested. An analysis showed that the entire market for computers consisted of only 20,000 machines per year and that Intel, as a latecomer, could hope to capture at most ten percent of that, far too small a business to fit into the strategy. In addition there was the problem of support. To sell a $ 10,000 computer one could afford to spent a few hundred dollars showing the customer how to use it. But for a $ 100 integrated circuit, which was every bit as difficult to use, such a teaching effort was out of the question [9].

Hoff was not convinced by the arguments. He had grasped the possibilities of a truly inexpensive computer, but the chances for his dream looked bleak. The chip was finished early in 1971 and was ready to proceed into production for one customer only. But then Hoff learned that the customer was asking for the price of the chip to be decreased, which meant that the entire contract was open for renegotiation. At the same time a new director of marketing joined Intel and Hoff launched a renewed campaign. This time he succeeded; Intel obtained a release and announced the 4004 microprocessor. It was to mark a new era for integrated circuits and for computers. At the time of its announcement there were some 70,000 computers in use in the world. During the next ten years this number would shoot up to four million [9].

At Intel Bob Noyce reached his zenith of fame. Everybody looked up to him; he was the all-American inventor, scientist and businessman, and model of rectitude. But he also had some flaws. He smoked incessantly, which no doubt contributed to his early death. And he had an affair (an open secret in Silicon Valley), which wrecked his marriage. He never *ordered* his employees to do something, he *convinced* them that it was the right thing to do; this approach was ideal while Fairchild and Intel were small, but didn't work well as the companies grew large.

In the 1980s the American semiconductor industry began losing ground to the Japanese. Noyce became active in getting the Japanese to open their internal market and stop their dumping (selling below cost) of semiconductors in the U.S. He helped form Sematech, an organization headquartered in

Austin, Texas, and funded by both government and industry to do basic research accessible to all U.S. participants. But the organization couldn't find a general manager, so Noyce reluctantly stepped in and moved to Austin [15]. At this stage in life he could have done anything he wanted (he was worth $ 147 million, even after the divorce), but he felt it was his duty.

It was an impossible job and Noyce hated every minute of it. Most scientists and engineers were on loan from their companies and had been told by their legal department not to divulge any secrets but to get as much information as possible. Little was accomplished and after 18 months Noyce began to look exhausted [15]. On June 3, 1990 he had a massive heart attack and died. He was 62 years old.

For Noyce's obituary the San Jose Mercury News, the main Silicon Valley newspaper, mistakenly used Kilby's photograph.

The quiet, unassuming Gordon Moore, who never strove to be number one, became one of the 400 richest men in America, even after having given half of his Intel stock to charity.

Despite its flaws the semiconductor industry is one of the most remarkable phenomena to hit the business world. Ever since Fairchild inadvertently created a host of spin-offs, competition has been fierce, almost deadly. Prices have fallen strikingly every year while the performance of the products leapt ahead without pause. Stand still for a few months and you are out.

When Fairchild Semiconductor was founded in 1957, world-wide sales of the industry amounted to $ 7 million and consisted of 3.5 million transistors. The year Intel came out with its first product (1969) sales passed the $ 250 million mark and about billion transistors were produced. Now (2006) sales amount to $ 250 billion and the number of transistors manufactured every year (mostly inside integrated circuits) is in the neighborhood of 10^{20}, or 10 billion for every man, woman and child in the world.

This growth came in spurts, with severe recessions in between. Each cycle was the same: expansion at an unsustainable rate, which created shortages; to increase their chance of getting products, customers placed the same order with several suppliers; as soon as the industry caught up with demand, multiple orders were canceled, thus sales plummeted and a recession followed.

It was Gordon Moore who observed early on that the complexity of integrated circuits seemed to multiply at a constant rate, regardless of how complex they already were, which became known as "Moore's Law". The rate didn't exactly turn out to be constant, it has moved from doubling every year to doubling every two years and now to doubling every 18 months. But the point is this: the complexity keeps on growing at a horrendous rate. (Moore, who has a knack for these things, also figured that sometime in 1997 the number of transistors produced exceeded the number of ants in the world).

Behind these figures are developments in technology that are truly astonishing. Jean Hoerni's planar process introduced photographic techniques to create the patterns for transistors, which enabled him to make their dimensions considerably smaller than anything produced before. By 1969 he and his colleagues achieved photographic resolution beyond that of photographs, creating devices with dimensions as small as 10um (about the thickness of paper). They now realized that there was a double incentive to making devices even smaller: not only could you cram more of them onto a chip at the same cost, but smaller devices were also faster. And so the race continued, down to 5um, 2um, 1um, 0.5um.

The progress in small dimensions is best illustrated with Intel's microprocessor chips. The first one, announced in 1971 had 2300 transistors and ran at 740kHz. In 1982 this increased to 134,000 transistors and 12MHz. The Pentium, announced in 1993 had 3.1 million transistors and ran at 66MHz, followed by 28 million transistors and 600 MHz in 1999. The latest computer chip (2005) contains 230 million transistors and runs at 3,400 MHz, or 3.4 GHz. (And anyone reading this 10 years from now will undoubtedly think how puny and slow computers were in 2005).

As dimensions were shrinking a barrier loomed, which appeared to be insurmountable: the wavelength of light. Visible light ranges in wavelength from 0.7um for red to 0.4um for violet. Producing features smaller than the wavelength of the light used to photograph it seemed impossible. But there was really no need to use light, which could be seen by the human eye. Above the visible spectrum there is ultra-violet, with a wave-length as short a 0.01um, or 10 nano-meters (10nm) and x-rays beyond that. With these (and many other) measures the dimensions on an integrated circuit have now shrunk to 65nm, which the operators can actually no longer see.

In parallel to shrinking dimensions wafer sizes have increased constantly to process more chips simultaneously and thus reduce cost. This is a remarkable development all by itself. A wafer is cut from a rod, which is slowly pulled out of molten silicon. The entire rod needs to be a perfect single crystal. Even when these rods had a diameter of only 25mm (1 inch), they were the largest crystals in existence; nature had never produced a crystal this big or this pure. Then we learned how to obtain the same quality with

100mm (4-inch), 200mm (8-inch) and now 300mm (12-inch) wafers.

As you make wafer sizes larger and device dimensions smaller, the requirement for cleanliness becomes extreme. Even the smallest dust particle in the path of the light can kill a transistor that, in the case of a microprocessor or memory, assigns the entire integrated circuit to the trash-heap. Ironically the human being turned out to be the biggest source of contamination; a breath after lunch can contaminate an entire wafer. That is why you see operators wearing space suits.

All this progress didn't come cheap. The cost of a wafer fab, about $ 30 million in 1969, has increased to $ 3 billion.

But the result speaks for itself: In a 1GB RAM, which contains 8 billion transistors, the cost per transistor has sunk to less than one-millionth of a cent.

12. Parting Shots

This book has been unfair to a considerable number of people. Besides the ones mentioned, there were many others who made significant discoveries or invented important things. At least several hundred; more likely several thousand.

And herein lies the problem: If they had all been represented fairly, your eyes would have glazed over and your knees would have buckled under the weight of the book. It simply can't be done.

We must keep in mind that the vast majority of inventions and discoveries consists of small steps, which occasionally trigger a larger one. Scientific progress is not primarily the fruit of a few extraordinary thinkers but the contribution of many people; people who will never get a Nobel Prize or become famous, but have the immeasurable satisfaction of having been there first. Quite often the distinction of having discovered or invented something important is simply a matter of luck.

And this army of unsung inventors and discoverers is ever restless. Their main driving force at the moment is the integrated circuit with its ever shrinking devices. At present it looks like integrated circuits will keep on becoming faster and more complex for another ten or twenty years, but eventually there will be limitation. For one thing, we are approaching the size of the atom and nobody has yet figured out how to make a semiconductor work with less than an atom.

The second looming limitation is cost. The expenditure for a small-dimension fabrication line is slowly approaching the gross national product of a small nation. At some point it will simply no longer be economical to make dimensions any smaller.

But then again there are thousand of people thinking about how to beat limitations and the solution might come from an entirely unexpected direction.

Is it all worth it? Are we better off now than we were earlier? Does technology improve our lives?

You are tempted to shout a resounding NO when the person next to you talks loudly on a cell-phone, when music is amplified to the level of pain, or when your computer tells you that it has to shut down now and should it send a report to Microsoft? In moments like these you wish for simpler times and electronics had never been invented.

Suppose we could freeze progress at any one point in time. Which time would you choose? Before the integrated circuit was invented? Fine, let's do that. Now there are no more personal computers. But there is also no internet or satellite communication, no VCRs or DVD players. Your car consumes twice a much fuel because there is no electronics to fine-tune it. Most of the medicine developed in the last 50 years doesn't exist because the scientific instruments needed for their development don't exist. There are no CAT scanners, MRI, or any other medical diagnostic tools. The list is almost endless.

The good clearly outweighs the bad by a wide margin.

References

Chapter 1: Almost Nothing
1   Benjamin, Park: A History of Electricity, John Wiley, New York, 1898
2   Farrington, Benjamin: Greek Science, Penguin Books, Harmondsworth, England, 1949
3   Gilbert, William: On the Lodestone, 1600 (in: Great Books of the Western World, Encyclopaedia Britannica)
4   Gray, Stephen: Experiments with Conductors and Non-Conductors (in Boynton, Holmes, ed.: The Beginning of Modern Science, Walter J. Black, Roslyn, New York, 1948)
5   Guericke, Otto von: Experimenta Nova Magdeburgica, Amsterdam, 1672
6   Heilbron, J. L.: Electricity in the 17th and 16th Centuries, University of California Press, Berkeley, 1979
7   Mottelay, Paul F.: Bibliographical History of Electricity and Magnetism, Charles Griffin, London, 1922
8   Rice, Patty C.: Amber, Van Nostrand Reinhold, New York, 1960
9   Taylor, Julie: Muslims in Medieval Italy, Lexington Books, 2003
10 Thompson, Silvanus P.: Petrus Peregrinus de Maricourt and his Epistola de Magnete, Proceedings of the British Academy, 1905/1906, pp. 376-408

Note: There were many additional contributors to the understanding of electricity and magnetism between 1550 and 1745. For a much more detailed account see Heilbron, above.

Chapter 2: Benjamin Franklin, Electrician
1   Benjamin, Park: A History of Electricity, John Wiley, New York, 1898
2   Cohen, I. Bernard: Benjamin Franklin's Experiments, Harvard University Press, Cambridge, Mass., 1941
3   Cohen, I. Bernard: Benjamin Franklin, Charles Scribner's Sons, New York, 1975
4   Cohen, I. Bernard: The Twohundreth Anniversary of Benjamin Franklin's Two Lightning Rod Experiments and the Introduction of the Lightning Rod, Proceedings of the American Philosophical Society, June 1952, pp. 331-366
5   Cohen, I. Bernard: Benjamin Franklin and the Mysterious "Dr. Spence", Journal of the Franklin Institute, January 1943, pp. 1-25
6   Cohen, I. Bernard: Prejudice Against the Introduction of Lightning Rods, Journal of the Franklin Institute, May 1952, pp. 393-440
7   Dibner, Bern: Benjamin Franklin, Electrician, Burndy Library, Norwalk, Conn., 1976
8   Dibner, Bern: Early Electrical Machines, Burndy Library, Norwalk, Corn., 1957
9   Finn, Bernard S.: Franklin as Electrician, Proceedings of the IEEE, Sept. 1976, pp. 1270-1273
10 Heilbron, J. L.: Electricity in the 17th and 18th Centuries, University of California Press, Berkeley, 1979
11 Home, R.W.: Franklin's Electrical Atmospheres, British Journal for the History of Science, 1972, pp. 131-151
12 Lemay, Leo J.: Ebenezer Kinnersley, University of Pennsylvania Press, Philadelphia, 1964
13 Lopez, Claude-Anne and Herbert, Eugenia W.: The Private Franklin, W. W. Norton, New York, 1975
14 Mackay, Charles: Extraordinary Popular Delusions and the Madness of Crowds, London 1841 (Farrar, Strauss & Giroux, New York, 1932)

15   Schonland, B.F.: The Flight of the Thunderbolts, Oxford, 1950
16   Van Doren, Carl: Benjamin Franklin, Viking Press, New York, 1939
17   Van Doren, Carl (ed.): Benjamin Franklin's Autobiographical Writings, Viking Press, 1945
18   White, Andrew D.: A History of the Warfare of Science with Theology in Christendom, 1895 (George Braziller, New York, 1955

Note: In addition to the people mentioned in this chapter several others proposed theories of electricity. For more detail see Heilbron, above.

Chapter 3: Nine Lives, Nine Discoveries

1   Agassi, Joseph: Faraday as a Natural Philosopher, University of Chicago Press, Chicago, 1971
2   Arago, Francois: Eulogy on Ampere, Smithsonian Institution Annual Report, 1872, pp. 111-171
3   Aykroyd, W.R.: Three Philosophers (Cavendish), William Heinemann, London, 1935
4   Berry, A. J.: Henry Cavendish, Hutchinson, London, 1960
5   Cavendish, Henry: Electrical Researches, edited by James Clerk Maxwell, Cambridge University Press, Cambridge, 1879
6   Coulson, Thomas: Joseph Henry, Princeton University Press, Princeton, 1950
7   Crowther, J. G.: Scientists of the Industrial Revolution (Cavendish), Cresset Press, London, 1962
8   Dibner, Bern: Alessandro Volta, Franklin Watts, New York, 1964
9   Dibner, Bern: Luigi Galvani, Burndy Library, Norwalk, Conn., 1971
10   Dibner, Bern: Oersted and the Discovery of Electromagnetism, Blaisdell Publ. Co., New York, 1962
11   Galvani, Luigi: Commentary on the Effects of Electricity on Muscular Motion (English Translation), Burndy Library, Norwalk, Conn., 1953
12   Geddes, L.A. arid Hoff, H.E.: The Discovery of Bioelectricity and Current Electricity, IEEE. Spectrum, Dec. 1971, pp. 38-46
13   Gillmor, C. Stewart: Coulomb and the Evolution of Physics and Engineering in 18th Century France, Princeton University Press, Princeton, 1971
14   Faraday, Michael: Historical Sketch of Electro-Magnetism, Annals of Philosophy, 1822, pp. 107-121
15   Faraday, Michael: Experimental Researches in Electricity (in Great Books of the Western World, Encyclopaedia Britannica)
16   Hamilton, James: A Life of Discovery (Faraday), Random House, New York, 2002
17   Heilbron, J. L.: Electricity in the 17th and 18th Centuries, University of California Press, Berkeley, 1979
18   Heimann, P.M.: Faraday's Theories of Matter and Electricity, British Journal for the History of Science, 1971, pp. 235-257
19   Henry, Joseph: The Papers of Joseph Henry, volumes 1-4, Smithsonian Institution Press, Washington, DC, 1972 - 1979
20   Hofmann, James E.: Andrè-Marie Ampere, Cambridge University Press, 1996
21   Landau, David and Charlotte B.: Andre-Marie Ampere, The Technology Review, Dec. 1948, pp. 3-10
22   Launay, Louis de: Le Grand Ampere, Librairie Academique Perrin, Paris, 1925
23   Molella, Arthur P.: The Electric Motor, the Telegraph, and Joseph Henry's Theory of Technological Progress, Proceedings of the IEEE, September 1976, pp. 1273-1278
24   Moyer, Albert.: Joseph Henry, Smithsonian Institution, 1997

25   Oersted, Hans C.: The Soul in Nature, London, 1852 (Dawsons, London, 1966)
26   Ohm Georg S.: Die Galvanische Kette mathematisch bearbeitet, Berlin, 1827 (The
      Galvanic Circuit Investigated Mathematically, Van Nostrand, New York, 1891)
27   Pancaldi, Giuliano: Volta, Princeton University Press, 2003
28   Pera, Marcello: The Ambiguous Frog, Princeton University Press, 1992
29   Schagrin, Morton L.: Resistance to Ohm's Law, Journal of Physics, 1963, pp. 536-547
30   Schimank, Hans: Georg Simon Ohm, Elektrotechnische Zeitschrift, March 16. 1939, pp.
      330-332
31   Stauffer, R. C.: Persistent Errors Regarding Oersted's Discovery of Electromagnetism,
      Isis, vol. 44,1953, pp. 307-310
32   Stauffer, R. C.: Speculation and Experiment in the Background of Oersted's Discovery of
      Electromagnetism, Isis, vol.44,1953, pp. 33-50
33   Thompson, Silvanus P.: Michael Faraday, Cassell, London, 1901
34   Tricker, R. A. R.: Early Electrodynamics, Pergamon Press, Oxford, 1965
35   Tricker, R. A. R.: The Contributions of Faraday & Maxwell to Electrical Science,
      Pergamon Press, Oxford, 1966
36   Turner, R. Steven: The Ohm-Seebeck Dispute, British Journal for the History of Science,
      1977, pp. 1-24
37   Visan, Tancrede de: La Vie Passionee de Andre-Marie Ampere, Agance Archat, Paris,
      1936
38   Volta, Alessandro, Philosophical Transactions of the Royal Society, 1893, pp. 285-291
39   Volta, Alessandro: On the electricity excited by the mere contact of conducting
      substances of different kinds, Philosophical Transactions of the Royal Society, (90) 1800,
      pp. 403-431, 744-746
40   Williams, L. Pearce: Ampere's Electrodynamic Molecular Model, Contemporary Physics,
      1962, 4, pp.113-123
41   Williams, L. Pearce: Michael Faraday, Basic Books Inc., New York, 1965
42   Williams, L. Pearce: What were Ampere's Earliest Discoveries in Electrodynamics?, Isis,
      1983, pp. 492-528
43   Wilson, George: The Life of the Hon. Henry Cavendish, Cavendish Society, London,
      1851
44   Zenneck, J.: Ohm, Elektrotechnische Zeitschrift, April 13. 1939, pp. 441-447

Chapter 4: A Very Big Ship
1   Bright, Charles: Submarine Telegraphs, Crosby Lockwood & Sons, London, 1898
      (reprint Arno Press, New York, 1974)
2   Carter, Samuel: Cyrus Field, G.P. Putnam & Sons, New York, 1968
3   Clarke, Arthur C.: Voice Across the Sea, William Luscombe, London, 1958
4   Coates, Vary T. and Finn, Bernard: A Retrospective Technology Assessment: Submarine
      Telegraphy, San Francisco Press, San Francisco, 1979
5   Dibner, Bern: The Atlantic Cable, Burndy Library, Norwalk, Con, 1959
6   Dugan, James: The Great Iron Ship, Harper & Brothers, New York, 1953
7   Fahie, J.J.: A History of Electric Telegraphy to the Year 1837, E. & F.N. Spon, London,
      188:4 (reprint Arno Press, New York, 1974)
8   Hemming, Richard: Die Entwicklung der Telegraphie und Telephonie, Verlag Johann
      Ambrosius Barth, Leipzig, 1908
9   Kieve, Jeffrey L.: The Electric Telegraph, David & Charles, Newton Abbot, England,
      1973
10   Larkin, Oliver W.: Samuel F.B Morse and American Democratic Art, Little, Brown &
      Co., Boston, 1954

11   Mabee, Carlton: The American Leonardo, the Life of Samuel F.B. Morse, Alfred A. Knopf, New York, 1943

12   Morse, Edward L.: Samuel F. B. Morse, His Letters and Journals, (1914), Da Capo Press, New York, 1973

13   Prime, Samuel I.: The Life of Samuel F.B. Morse, D. Appleton & Co., New York, 1:375 (reprint Arno Press, New York, 1974)

14   Rolt, L.T.C.: Isambard Kingdom Brunel, Longmans Green & Co., London, 1957

15   Russell, W.H.: The Atlantic Telegraph, (1865, David & Charles, Newton Abbot, England, 1972

16   Shiers, George (ed.): The Electric Telegraph, Arno Press, New York, 1977

17   Smith, Willoughby: The Rise and Extension of Submarine Telegraphy, J.S. Virtue & Co, London, 1891 (reprint Arno Press, New York, 1974)

18   Thompson, Silvanus P.: The Life of Lord Kelvin, (1910), Chelsea Publishing Co., New York, 1976.

Chapter 5: Mr. Watson, Come Here!

1   Bell, Alexander G.: The Bell Telephone, American Bell Telephone Co., Boston, 1908 (reprint Arno Press, New York, 1974)

2   Boettinger, H.M.: The Telephone Book, Riverwood Publishers, Croton-on-Hudson, New York, 1977

3   Brooks, John: Telephone, the First Hundred Years, Harper $ Row, New York, 1976

4   Bruce, Robert V.: Bell, Alexander Graham Bell and the Conquest of Solitude, Little, Brown & Co., Boston, 1973

5   Fagen, M.D. (ed.): A History of Engineering arid Science in the Bell System, The Early Years (187-1925, Bell Telephone Laboratories, 1975

6   Goulden, Joseph: Monopoly, G.P. Putnam's sons, New York, 1968

7   Hoddeson, Lillian: The Emergence of Basic Research in the Bell Telephone System, 1875-1915, Technology and Culture, 1981, pp. 512-544

8   Hounshell, David A.: Elisha Gray and the Telephone, Technology and Culture, 1975, pp. 133-161

9   Hounshell, David A.: Bell and Gray: Contrast in Style, Politics, and Etiquette, Proceedings of
the IEEE, September 1976, pp. 1305-1314

10   Langdon, William C.: Myths of Telephone History, Bell Telephone Quarterly, April 1933, pp. 123-140

11   Paine, Albert B.: In One Man's Life (Theodore N. Vail), Harper & Brothers, New York, 1921

12   Schiavo, Giovanni E.: Antonio Meucci, The Vigo Press, New York, 1958

13   Thompson, Silvanus P.: Philipp Reis, Inventor of the Telephone, Spon, London, 1883

14   Watson, Thomas A.: Exploring Life, D. Appleton & Co., New York, 1926

15   Watson, Thomas A.: The Birth and Babyhood of the Telephone, American Telephone and Telegraph Co., 1938

Chapter 6: Tesla

1   Anderson, John M. and Saby, John S.: The Electric Lamp: 100 Years of Applied Physics, Physics Today, October 1979, pp. 32-40

2   Barham, G. Basil: The Development of the Incandescent Lamp, Scott Greenwood, London, 1912

3   Bowers, Brian: A History of Electric Light and Power, Peter Peregrinus, Stevenage, UK, 1982

4   Brittain, James E. (ed.): Turning Points in American Electrical History, IEEE Press, New York, 1977

5   Brown, C.N.: J.W. Swan and the Invention of the Incandescent Electric Lamp, Science Museum, London, 1978

6   Clark, Ronald W.: Edison, G.P. Putnam's Sons, New York, 1977

7   Conot, Robert: A Streak of Luck (Edison), Seaview Books, New York, 1979

8   Derganc, Christopher S.: Thomas Edison and his Electric Lighting System, IEEE Spectrum, February 1979, pp. 50-59

9   Dibner, Bern: James Clerk Maxwell, IEEE Spectrum, December 1964, 56

10   Everitt, C.W.: James Clerk Maxwell, Charles Scribner's Sons, New York, 1975

11   Friedlander, Gordon: Tesla: Eccentric Genius, IEEE Spectrum, June 1972, pp. 26-29

12   Hertz, Johanna (ed.): Heinrich Hertz, San Francisco Press, San Francisco, 1977

13   Jehl, Francis: Menlo Park Reminiscences, Edison Institute, Dearborn, Mich., 1936

14   Jonnes, Jill, Empires of Light, Random House, New York, 2003

15   Josephson, Matthew: Edison, McGraw-Hill, New York, 1959

16   Leupp, Francis E.: George Westinghouse, Little, Brown & Co., Boston, 1918

17   Marshall, Eliot: Seeking Redress for Nikola Tesla, Science, Oct. 30. 1981, pp.523-525

18   Martin, Thomas C.: The Inventions, Researches and Writings of Nikola Tesla, (1894), Omni Publications, Hawthorne, Ca., 1977

19   Maxwell, James C.: Treatise on Electricity and Magnetism, Oxford University Press, London, 1873

20   Morrison, Philip and Emily: Heinrich Hertz, Scientific American, Dec. 1957, pp. 98 - 104

21   O'Neill, John J.: Prodigal Genius, The Life of Nikola Tesla, (1944), Angriff Press, Hollywood, Ca., 1981

22   Pratt, Haraden: Nikola Tesla, 1856 - 1943, Proceedings of the IRE, Sept. 1956, pp. 1106 - 1108

23   Prout, Henry G.: A Life of George Westinghouse, The American Society of Mechanical Engineers, New York, 1921

24   Reynolds, Terry S. and Bernstein, Theodore: The Damnable Alternating Current, Proceedings of the IEEE, September 1976, pp 1339-1343

25   Seifer, Marc J.: The Life and Times of Nikola Tesla, Carol Publishing, Secaucus, N.J., 1996

26   Sweeney, Kenneth M.: Nikola Tesla, Science, May 16. 1958, pp. 1147-1159

27   Tesla, Nikola: My Inventions, Electrical Experimenter, May, June, July, October 1919

28   Tesla, Nikola: Lectures, Patents, Articles, Nikola Tesla Museum, Beograd, 1956

29   Tesla, Nikola, Colorado Springs Notes, 1899-1900, Nolit, Boegrad, Yugoslavia, 1978

30   Tesla, Nikola: Experiments with Alternate Currents, (1904), Publications, Hawthorne, Ca., 1979

Chapter 7: Revelation

1   Anderson, David L.: The Discovery of the Electron, D. Van Nostrand, Princeton, NJ, 1964

2   Dahl, Per F.: Flash of the Cathode Rays, IOP Publishing, Bristol, 1997

3   Dibner, Bern: The New Rays of Professor Roentgen, Burndy Library, Norwalk, Conn., 1963

4   French, A.P. and Kennedy, P.J. (ed.): Niels Bohr, Harvard University Press, 1985
5   Glasser, Otto: Wilhelm Conrad Roentgen, Julius Springer, Berlin, 1931
6   Glasser, Otto: The Human Side of Science, Cleveland Clinic Quarterly, July 1953, pp. 400-405
7   Heilbron, J.L.: Ernest Rutherford, Oxford University Press, 2003
8   Millikan, Robert A.: The Electron, University of Chicago Press, Chicago, 1917
9   Millikan, Robert A.: The Autobiography of Robert A. Millikan, Prentice-Hall, New York, 1950
10   Nitske, W. Robert: Roentgen, University of Arizona Press, 1971
11   Oliphant, Mark: Rutherford, Elsevier Publ. Co., Amsterdam, 1972
12   Sarton, George: The Discovery of X-Rays, Isis, 26, 1937, pp.347-369
13   Snow C.P.: The Physicists, Little, Brown, Boston, 1981
14   Snow, C.P.: Variety of Men, Macmillan, London, 1967
15   Thomson, George P.: J.J. Thomson and the Cavendish Laboratory, Doubleday, Garden City, NY, 1965
16   Thomson, J.J.: Recollections and Reflections, Macmillan, New York, 1937

Chapter 8: Armstrong
1   Aitken, Hugh G.: Syntony and Spark, John Wiley, New York, 1976
2   Appleyard, Rollo: Pioneers of Electrical Communication, Macmillan, London, 1930
3   Archer, Gleason L.: History of Radio, The American Historical Society, New York, 1938
4   Archer, Gleason L.: Big Business and Radio, The American Historical Company, New York, 1939
5   Arvin, W.B.: The Life and Work of Lee de Forest, Radio News, Oct. 1924 to September 1925
6   Baker, W.J.: The History of the Marconi Company, St. Martin's Press, New York, 1971
7   Blanchard, Julian: The History of Electrical Resonance, Bell System Technical Journal, Oct. 1941, pp. 415 - 433
8   Blake, G.G.: History of Radio Telegraphy and Telephony, Radio Press, London, 1926
9   Briggs, Asa: The Birth of Broadcasting, Oxford University Press, London, 196
10   Carneal, Georgette: A Conqueror of Space (de Forest), Horace Liveright, New York, 1930
11   Chipman, Robert A.: De Forest and the Triode Detector, Scientific American, March 1965, pp. 92 - 100
12   De Forest, Lee: Father of Radio, Wilcox & Follet, Chicago, 1950
13   Douglas, Alan: The Crystal Detector, IEEE Spectrum, April 1981, pp. 64 - 67
14   Erickson, D.H.: Armstrong's Fight for FM Broadcasting, University of Alabama Press, 1973
15   Fahie, J.J.: A History of Wireless Telegraphy (1901) (Arno Press, New York, 1971)
16   Fleming, J.A.: The Thermionic Valve, The Wireless Press, London, 1919
17   Fleming, J.A.: Memories of a Scientific Life, Marshall, Morgan & Scott, London, 1934
18   Fortune Magazine: Armstrong of Radio, Feb. 1948 (also Oct. 1939)
19   Harrison, Arthur P.: Single Control Tuning, Technology and Culture, April 1979, pp. 296-321

20 Hawks, Ellison: Pioneers of Wireless, Methuen, London, 1927 (Arno Press, New York, 1974)

21 Howe, G.W.: The Genesis of the Thermionic Valve, in Thermionic Valves, 1904-1954, The Institution of Electrical Engineers, London, 1955

22 Jolly, W.P.: Marconi, Constable, London, 1972

23 Leinwoll, Stanley: From Spark to Satellite, Charles Scribner's Sons, New York, 1979

24 Lessing, Lawrence: Man of High Fidelity: Edwin Howard Armstrong, Lippincott, Philadelphia, 1956

25 Lewis, Tom: Empire of the Air, Harper Collins, New York, 1991

26 Lubell, Samuel: Magnificient Failure (de Forest), Saturday Evening Post, Jan. 17., 24., 31., 1942

27 Lyons, Eugene: David Sarnoff, Harper & Row, New York, 1966

28 Maclaurin, W. Rupert: Invention and Innovation in the Radio Industry, Macmillan, New York, 1949 (Arno Press, 1971)

29 Marconi, Degna: My Father, Marconi, McGraw Hill, New York, 1962

30 Marriott, Robert H.: Radio Ancestors, IEEE Spectrum, June 1966, pp. 52-61

31 O'Dea, W.T.: Radio Communications, Science Museum, London, 1934

32 Phillips, Vivian J.: Early Radio Wave Detectors, Peter Peregrinus, Stevenage, UK, 1980

33 Schueler, Donald G.: Inventor Marconi: brilliant, dapper, tough to live with, Smithsonian, March 1982, pp. 126 - 147

34 Shiers, George (ed.): The Development of Wireless to 1920, Arno Press, New York, 1977

35 Shiers, George: The First Electron Tube, Scientific American, March 1969, pp. 104 - 112

36 Suesskind, Charles: Distortions in the History of Radio, Proceedings of the IEEE, Febr. 1965, pp. 162-164

37 Tyne, Gerald F.: Saga of the Vacuum Tube, Howard W. Sams, Indianapolis, 1977

Chapter 9: Farnsworth

1 Abramson, Albert: Pioneers of Television, SMPTE Journal, July 1981, pp. 578-590

2 Abramson, Albert: Electronic Motion Pictures, Arno Press, New York, 1974

3 Dobriner, Ralph: Vladimir Zworykin, Electronic Design, September 1. 1977, pp. 112-115

4 Dreher, Carl: Sarnoff, New York Times Book Co, New York, 1977

5 Everson, George: The Story of Television, W.W. Norton & Co., New York, 1949

6 Exwood, Maurice: John Logie Baird, Institution of Electronic and Radio Engineers, London, 1976

7 Farnsworth, Elma: Distant Vision, PemperlyKent, Salt Lake City, 1989

8 Fink, Donald: Television Broadcasting in the United States, 1927-1950, Proceedings of the IRE, February 1951, pp. 116-123

9 Fink, Donald: Perspectives on Television, Proceedings of the IEEE, Sept. 1976, pp. 1322-1338

10 Fisher, David E. and Fisher, Jon: Tube, Harcourt Brace, 1997

11 Garratt, G.R. and Mumford, A.H.: The History of Television, Proc. IEE, Apr./May 1952, pp.25-42

12 Godfrey, Donald G.: Philo T. Farnsworth, Utah Press, 2001

13   Hogan, John V.: The Early Days of Television, SMPTE Journal, November 1954, pp.169-173
14   Jenkins, C. Francis: Radio Vision, Proceedings of the IRE, Nov. 1927, pp 958-964
15   Jensen, A.G.: The Evolution of Modern Television, SMPTE Journal, Nov. 1963, pp.174-188
16   Lyons, Eugene: David Sarnoff, Harper & Row, New York, 1966
17   McGee, J.D.: Distant Electric Vision, Proceedings of the IRE, June 1950, pp.596-608
18   Moseley, Sidney: John Baird, Odhams Press, London, 1952
19   Percy, J.D.: John L. Baird, London, 1950
20   Schatzkin, Paul and Kiger, Bob: Philo T. Farnsworth, Televisions, vol.5 (1979), no. 1-4
21   Schwartz, Evan I.: The Last Lone Inventor, Perennial, 2002
22   Shiers, George: Early Schemes for Television, IEEE Spectrum, May 1970, pp. 24-34
23   Shiers, George: Television Fifty Years Ago, Television, Jan./Feb. 1976, pp. 6-10, 13
24   Shiers, George: Historical Notes on Television Before 1900, SMPTE Journal, March 1977, pp. 129-137
25   Shiers, George (ed.): Technical Development of Television, Arno Press, New York, 1977
26   Shiers, George: The Rise of Mechanical Television, 1901-1930, SMPTE Journal, June 1981, pp. 508-521
27   Udelson, Joseph H.: The Great Television Race, University of Alabama Press, 1982
28   Waldrop, Frank C. and Borkin, Joseph: Television, William Morrow & Co., New York, 1938
29   Zworykin, Vladimir K.: The Early Days, Television Quarterly, Nov. 1962, pp.69-73

Chapter 10: A 30-Ton Brain

1   Allgemeine Deutsche Biographie: Schickard, Wilhelm S.
2   Austrian, Geoffrey D.: Herman Hollerith, Columbia University Press, New York, 1982
3   Belden, Thomas G. and Marva R.: The Lengthening Shadow, Little, Brown & Co., Boston, 1962
4   Bowden, B.V. (ed.): Faster than Thought, Sir Isaac Pitman & Sons, London, 1953
5   Ceruzzi, Paul E.: Reckoners, Greenwood Press, Westport, CT, 1983
6   Chapman, S.: Blaise Pascal, Nature, Oct. 31. 1942, pp.508-509
7   Crowthier, J.G.: Scientific Types (Charles Babbage), Barrie & Rockliff, London, 1968
8   Evans, Christopher: Pioneers of Computing, Cassette tapes, Science Museum, London, 1975/76
9   Fishman, Katharine D.: The Computer Establishment, Harper & Row, New York, 1981
10   Freedman, Russell and Morriss, James E.: The Brains of Animals and Man, Holiday House, New York, 1972
11   Freytag Loeringhoff, B.: Ueber die erste Rechenmaschine, Physikalische Blaetter, 1958, #8, pp.361-365
12   Gardner, W. David: Will the Inventor of the First Computer Please Stand Up?, Datamation, February 1974, pp.84, 88-90
13   Goldstine, Herman H.: The Computer from Pascal to von Neumann, Princeton University Press, 1972

14   Haugeland, John: Mind Design, MIT Press, Cambridge, Mass., 1981
15   Hawkins, Jeff: On Intelligence, Times Books, Henry Hall & Co., New York, 2004
16   Heims, Steve J.: John von Neumann and Norbert Wiener, MIT Press, Cambridge, Mass., 1981
17   Hubel, David H.: The Brain, Scientific American, September 1979, pp. 44-53
18   Hyman, Antony: Charles Babbage, Oxford University Press, Oxford, 1982
19   Kandel, Eric R.: In Search of Memory, W.W. Norton & Co., New York, 2006
20   Kean, David W.: The Computer and the Countess, Datamation, May 1973, pp.60-63
21   Kreiling, Frederick C.: Leibniz, Scientific American, May 1968, pp.94-100
22   Kurzweil, Ray, The Age of Spiritual Machines, Viking Penguin, New York, 1999
23   Lukoff, Herman: From Dits to Bits, Robotics Press, Portland, Oregon, 1979
24   Martin, T.C.: Counting a Nation by Electricity, The Electrical Engineer, November 11. 1891, pp.521-530
25   McCorduck, Pamela: Machines Who Think, W.H. Freeman, San Francisco, 1979
26   Metropolis et al (ed.): A History of Computing in the Twentieth Century, Academic Press, New York, 1980
27   Moseley, Maboth: Irascible Genius (Charles Babbage), Henry Regnery, Chicago, 1964
28   Morrison, Philip and Emily: The Strange Life of Charles Babbage, Scientific American, April 1952, pp.66-73
29   Randell, Brian: The Origins of Digital Computers, Springer Verlag, Berlin/New York, 1973
30   Rodgers, William: Think, A Biography of the Watsons and IBM, Stein and Day, New York, 1969
31   Rose, Steven: The Conscious Brain, Vintage Books, New York, 1976
32   Serrell, R. et al: The Evolution of Computing Machines and Systems, Proceedings of the IRE, May 1962, pp.   1039-1058
33   Sobel, Robert: IBM, Times Books, 1981
34   Stern, Nancy: From Eniac to Univac, Digital Press, Bedford, Mass., 1981, also IEEE Spectrum, December   1981, pp.   61-69; also The Eckert-Mauchly Computers, Technology and Culture, Oct. 1982, pp. 569-582
35   Stevens, Leonard A.: Explorers of the Brain, Alfred A. Knopf, New York, 1971
36   Tomayko, James E.: The Stored-Program Concept, Technology and Culture, Oct. 1983, pp. 660-663
37   Tropp, Henry: The Effervescent Years: A Retrospective, IEEE Spectrum, Febr. 1974, pp. 70-79
38   Wittrock, M.C. et al: The Human Brain, Prentice-Hall, Englewood Cliffs, N.J., 1977
39   Wooldridge, Dean E.: The Machinery of Life, McGraw-Hill, New York, 1966
40   Wooldridge, Dean E.: Mechanical Man, McGraw-Hill, New York, 1968

Chapter 11: Noyce

1   Bardeen, J. and Brattain, W.H.: The Transistor, a Semiconductor Triode, Physical Review, July 15, 1948, pp.230-231
2   Berlin, Leslie: The Man Behind the Microchip (Noyce), Oxford University Press, 2005
3   Bottom, Virgil E.: Invention of the Solid-State Amplifier, Physics Today, February 1964, pp.24-26
4   Brattain, Walter H.: Genesis of the Transistor, The Physics Teacher, March 1968, pp.109-114
5   Braun, Ernest and MacDonald, Stuart: Revolution in Miniature, Cambridge University Press,
Cambridge, 1978

6 Bylinsky, Gene: The Innovation Millionaires, Charles Scribner's Sons, New York, 1976

7 Electronics: The Improbable Years, Electronics, February 19. 1963, pp.78-90

8 Hanson, Dirk: The New Alchemists, Little, Brown & Co., Boston, 1982

9 Interviews with Phil Ferguson, Victor Grinich, Jean Hoerni, Eugene Kleiner, Robert Noyce and Ted Hoff.

10 Kilby, Jack: Invention of the Integrated Circuit, IEEE Transactions on Electron Devices, July 1976, pp. 648-654

11 Kurylo, Friedrich and Susskind, Charles: Ferdinand Braun, MIT Press, 1981

12 Lécuyer, Christophe: Making Silicon Valley, MIT Press, Cambridge, Mass.

13 Lydon, James: The Semiconductor Industry, Electronic News, January 25 1982, pp.14-22

14 Noyce, Robert N.: Microelectronics, Scientific American, September 1977, pp. 63-69

15 Pearson, G.L. and Brattain, W.H.: History of Semiconductor Research, Proceedings of the IRE, December 1955, pp. 1794-1806

16 Reid, T.R.: The Chip, Simon and Schuster, New York, 1984

17 Shockley, W.: The Invention of the Transistor: An Example of Creative-Failure Methodology, National Bureau of Standards Publication # 3:38 (May 1974)

18 Shockley, William: The Path to the Conception of the Junction Transistor, IEEE Transactions on Electron Devices, July 1976, pp. 597-620

19 Teal, Gordon: Single Crystals of Germanium and Silicon - Basic to the Transistor and Integrated Circuit, IEEE Transactions on Electron Devices, July 1976, pp. 621-639

20 Walker, Rob: The Fairchild Chronicles, DVD of interviews, Panalta, Palo Alto, CA 2005

21 Weiner, Charles: How the Transistor Emerged, IEEE Spectrum, January 1973, pp. 24-33

22 Wolff, Michael F.: The Genesis of the Integrated Circuit, IEEE Spectrum, August 1976, pp. 45-53

Index

Picture Credits

William Gilbert: Huntington Library, San Marino, CA
Otto von Guericke: Deutsches Museum, Munich, Germany
Pieter van Musschenbroek: Deutsches Museum, Munich, Germany
Benjamin Franklin: Historical Society of Pennsylvania, Philadelphia, PA
Charles Coulomb: Emilio Segrè Visual Archives, American Institute of Physics, New York, NY
Luigi Galvani: Science Museum, London, UK
Alessandro Volta: The Institution of Electrical Engineers, London
Hans Christian Oersted: Science Museum, London, UK
André Marie Ampere: Musee Ampere, Polymieux au Mont-d'Or, France
Georg Ohm: Deutsches Museum, Munich, Germany
Joseph Henry: Smithsonian Institution, Washington, DC
Michael Faraday: Smithsonian Institution, Washington, DC
Samuel Morse: Science Museum, London, UK
Cyrus Field: Science Museum, London, UK
Lord Kelvin: Science Museum, London, UK
Great Eastern: Huntington Library, San Marino, CA
Alexander Graham Bell: Bell Laboratories / AT&T Archives, Warren, NJ
Thomas Watson: Bell Laboratories / AT&T Archives, Warren, NJ
Thomas Alva Edison: Edison National Historic Site, Orange, NJ
Nikola Tesla: Smithsonian Institution, Washington, DC
George Westinghouse: Westinghouse Historical Collection, Pittsburgh, PA
Heinrich Hertz: Deutsches Museum, Munich, Germany
Wilhelm Roentgen: Deutsches Roentgen Museum, Remscheid-Lennep, Germany
J.J. Thomson: Cavendish Laboratory, Cambridge University, Cambridge, UK
Robert Millikan: California Institute of Technology, Pasadena, CA
Ernest Rutherford: Cavendish Laboratory, Cambridge University, Cambridge, UK
Niels Bohr: Emilio Segrè Visual Archives, American Institute of Physics, New York, NY (
Guglielmo Marconi: Science Museum, London, UK
Lee de Forest: David Sarnoff Library, Princeton, NJ
David Sarnoff: Youssuf Karsh, David Sarnoff Library, Princeton, NJ
Howard Armstrong: Smithsonian Institution, Washington, DC
Philo Farnsworth: Elma Farnsworth, Salt Lake City, UT
Vladimir Zworykin: David Sarnoff Library, Princeton, NJ
John Bardeen: Bell Laboratories / AT&T Archives, Warren, NJ
Walter Brattain: Bell Laboratories / AT&T Archives, Warren, NJ
William Shockley: Alfred Eisenstaedt
Jean Hoerni: Jean Hoerni
Robert Noyce: Intel Corporation, Santa Clara, CA
Gordon Moore: Intel Corporation, Santa Clara, CA
Ted Hoff: Ted Hoff

Drawings: Bruce K. Hopkins, Berkeley, CA